USING SOCIAL MARKETING FOR PUBLIC EMERGENCY PREPAREDNESS

Less than half of the public in the United States have taken the three steps to prepare for emergencies that are recommended by the Federal Emergency Management Agency (FEMA) and the Red Cross: having a 3-day emergency kit, a family communication plan, and knowing where to get information during an emergency. Although emergency managers attempt to train the public, often they are only able to distribute brochures and make public notifications. For a variety of reasons, the public frequently ignores this guidance, leaving people more vulnerable during emergencies.

This book applies the process of social marketing, which has been used widely in public health and other disciplines, to the lack of public preparedness. Written for emergency managers in government and nonprofit agencies, students, and volunteers, the book provides enough background and resources to enable the user to carry out an effective emergency preparedness campaign in their community and maintain it over time. Unlike preparing one message for everyone, social marketing involves working with smaller communities to identify what and how people want to learn, training them, and then maintaining that relationship to ensure their preparedness. Because most emergency management agencies lack resources to take on such an initiative, the book provides readers with low cost methods to begin a social marketing program.

Nancy Meyer-Emerick is an Associate Professor of Public Administration at Cleveland State University (CSU). Prior to joining the CSU faculty in 1999, she worked in local government in Florida, primarily in public works and environmental health. Her research interests include organizational behavior, emergency preparedness, and environmental health.

How refreshing and encouraging to have an entire book focused on applying a social marketing approach to impact one major social issue. The book not only inspires; it also instructs. It doesn't just cover what to do to influence emergency preparedness in communities; it walks you through the steps to base unique and comprehensive strategies on unique priority audiences. In the end, readers will understand how to develop a strategic social marketing plan that will provide the highest return on investment of resources.

Nancy R. Lee, University of Washington and Social Marketing Services, Inc.

Nancy Meyer-Emerick provides an evidence-driven primer that outlines the core concepts and methodologies behind social marketing applications in emergency preparedness program management. From behavioral change models to social capital to program planning, this text outlines the fundamental components for leveraging modern messaging technologies to move the public from complacency to readiness. The content will be immediately applicable for any administrator interested in expanding their knowledge of the evolving and dynamic field of emergency management.

Lauren Ohl-Trlica, Arlington County Office of Emergency Management

Using Social Marketing for Public Emergency Preparedness tackles the real-world issues of blending the best of 'textbook' social marketing with the practical realities of applying it in communities. It provides an excellent guide and case study for any social marketer who wants to understand how to transform theory and evidence-based research in social marketing into real-life, practical solutions.

Craig Lefebvre, RTI International

USING SOCIAL MARKETING FOR PUBLIC EMERGENCY PREPAREDNESS

Social Change for Community Resilience

Nancy Meyer-Emerick

Routledge
Taylor & Francis Group
NEW YORK AND LONDON

First published 2016
by Routledge
605 Third Avenue, New York, NY 10017

and by Routledge
2 Park Square, Milton Park, Abingdon, Oxon OX14 4RN

Routledge is an imprint of the Taylor & Francis Group, an informa business

Library of Congress Cataloging-in-Publication Data
A catalog record for this book has been requested

ISBN: 978-0-765-64577-7 (hbk)
ISBN: 978-0-765-64578-4 (pbk)
ISBN: 978-1-315-68822-0 (ebk)

Typeset in Bembo
by Apex CoVantage, LLC

CONTENTS

INTRODUCTION

> Clearly, compared with information dissemination (e.g., pamphlets distributed to mail boxes), engagement based approaches to public outreach are costly in terms of time, commitment, resources and expertise required to develop and sustain the process. However, the benefits of doing so are substantial, particularly with regard to increasing the likelihood of developing sustained community involvement in risk management.
>
> *(Paton and McClure 2013, 142)*

Purpose of the Book

This book is designed to provide methods and resources to increase the effectiveness of public emergency preparedness training. It includes perspectives on behavioral and social change from several disciplines but focuses on one method, social marketing, that incorporates many of those theories. Social marketing has been used in public health training and other disciplines, but rarely in emergency preparedness. It is *not* marketing in the conventional sense; rather, it is a method of involving people in planning and carrying out changes that will improve their lives, not the bottom line of a corporation. It should also not be confused with social *media*, although that is one tool that may be used to help people to remember the positive changes they want to make.

Social marketing involves several phases. The following summary is based on Lee and Kotler's (2011) planning primer (40–41). The first step is a rich description of the background of the problem or innovation, the purpose of the social marketing effort, and people who will be the focus of the training. Next a situation analysis is done, typically by identifying strengths, weaknesses, opportunities, and threats (i.e., a SWOT analysis). Based on this information, the population is then segmented into priority or target groups based on, for example, who is at greatest risk, who is in greatest need, or who is most crucial for resolving the problem or achieving the social change. Behavioral objectives and goals are developed for these groups along with the barriers, benefits, competition, and outside influences they might experience. Then a positioning statement is developed to convey the benefits and value of the training for the priority group.

At this point, four "marketing strategies" (i.e., the 4Ps) are used but in an atypical way. The product is the benefit of the new behavior. The price is the cost of adopting the behavior. Place is a convenient site and promotion includes persuasive messages to encourage people. The design of follow-up and evaluation is determined as part of the overall plan and included in the budget. The roles and duties of everyone who is involved in the process are spelled out. Then the training is conducted to meet the aspirations and needs of the priority groups.

Unlike most training, social marketing includes an ongoing relationship between the trainers and priority groups to determine whether people have continued the new behaviors. If not, the training methods and messages can be revised and members of the focal group can assist in delivering them within their community. Think of how positive changes that individuals make, such as quitting smoking or driving sober, spill over into the larger community. That is the aim of social marketing: social change.

Lefebvre's (2013) definition captures the breadth and depth of this process:

> Social Marketing develops and applies marketing concepts and techniques to create value for individuals and society. This is done through the integration of research, evidence-based practice and social-behavioral theory together with the insights from individuals, influencers, and stakeholders. These inputs and perspectives are used to design more effective, efficient, sustainable and equitable approaches to enhance social wellbeing. The approach is one that encompasses all of the processes and outcomes that influence and are associated with change among: individuals, organizations, social networks and social norms, communities, businesses, markets, and public policy [Lefebvre 2012].
>
> (Lefebvre 2013, 502–503)

As anyone who has tried to change a habit knows, behavioral change is difficult. Even positive changes, like marriage or the arrival of a newborn, are stressful for people. Encouraging change requires expertise and resources beyond those typical of most emergency management agencies and nonprofits, such as the American Red Cross. Scientists have many good theories of what works, and I'll describe some of them in Chapter 2 as a prelude to diving into the social marketing methodology.

I discovered social marketing while researching how to train my local community to shelter-in-place and evacuate. Our Local Emergency Planning Committee

(LEPC) identified this need, but as we developed the plan, our team was told by a few first responders and members of the community that "you can't train the public." We routinely receive information about how to prepare, but only about one third of the U.S. population actually takes those actions.

I began to look for ways that we could increase the number of people who prepare by expanding on the Ready.gov and Red Cross messages that encourage people to "be informed, make a plan, build a kit and get involved." Managers and professionals in the emergency services community know that distributing information alone is not sufficient to motivate someone to change their behavior. Furthermore, while they may know the number of pieces of literature that were distributed to a specific group, they typically lack the resources to determine how the information was received and most importantly, whether people did the preparedness activities. Veil, Littlefield, and Rowan (2009) emphasize that research and evaluation are necessary to understand how segments of the population perceive and respond to specific hazards in order to craft effective messages and preparedness activities.

Increased Need for Preparedness

The level of public emergency preparedness has changed very little despite the September 11 attacks and the devastation caused by Hurricanes Katrina and Sandy and other storms and tornadoes. Due to climate change, natural emergencies and disasters are increasing more rapidly than anticipated just a few years ago. At the time of this writing, federal, state, and local governments are financially strapped due to their responses to storms, wildfires, droughts, and flooding. Without awareness and assistance from citizens who have learned to prepare and take care of themselves in emergencies, there will be greater injury and loss of life.

Although some emergency management professionals and first responders are against involving civilians in emergencies, the public's role, particularly in preparedness, is vital to developing a nationally resilient response to climate change. There are simply not enough professionals to manage large-scale incidents, particularly protracted ones along with long-term recovery efforts. It is possible to train the public because most people want to protect themselves and their families. With local resources and community support, public preparedness can become a reality. In the United States, people who live in areas that are prone to frequent natural disasters are usually more prepared than those who do not, as are those who know someone who has experienced a disaster. However, as will be discussed later in the book, changes in where we live and the arrival of climate change highlight the necessity of personal responsibility for emergency preparedness.

In their most recent report, *Climate Change 2014: Impacts, Adaptation, and Vulnerability*, the United Nations Intergovernmental Panel on Climate Change emphasized that "climate change is already having sweeping effects on every continent and throughout the world's oceans . . . and they warned that the problem is likely to grow substantially worse unless greenhouse emissions are brought

FIGURE INTRO 1.1 Climate change will affect us all (Image courtesy of darksite/iStock)

under control" (Gillis 2014). The report stated that "throughout the 21st century, climate-change impacts are projected to slow down economic growth, make poverty reduction more difficult, further erode food security, and prolong existing and create new poverty traps, the latter particularly in urban areas and emerging hotspots of hunger . . . the report also cites the possibility of violent conflict over land or other resources, to which climate change might contribute indirectly" (Gillis 2014).

The American Association for the Advancement of Science Climate Science Panel in their 2014 report, *What We Know: The Reality, Risks, and Response to Climate Change*, had three key messages for the American people: 1) climate change is happening here and now; 2) we risk "abrupt, unpredictable and potentially irreversible changes with highly damaging impacts"; 3) "the sooner we act, the lower the risk and cost. And there is much we can do" (Gillis 2014, 4). Their report concluded with encouragement but also a dire warning: "when scientists identified the grave environmental threats posed by the acid rain and the ozone hole, they worked together with other stakeholders . . . to develop solutions that would successfully reduce the threats while minimizing short- and long-term impacts . . . however, successfully responding to climate change will test our resolve and ingenuity in ways unlike any other environmental challenge we have faced" (Gillis 2014, 20).

Description of the Book

The book begins with a brief overview of current public preparedness in the United States and why it is important that we pay attention to the National Academies' recommendations for fostering national disaster resilience. Chapter 2 is a detailed description of methods currently used for public emergency preparedness training in order to set the stage for a more detailed explanation of social marketing. Chapter 3 is a case study from my own community, documenting formative research we used to identify the groups most ready to change, most at risk, and most critical for success of the social marketing effort. Chapter 4 documents the social marketing planning process using worksheets and information provided by the authors. Because tapping into social networks and social capital is very helpful when working with social marketing priority groups, Chapter 5 describes what they are and how to identify them. Chapter 6 provides alternative resources for local emergency managers and others who are faced with financial and staffing constraints. In Chapter 7, I discuss Rothschild's (1999) "continuum of interventions" and other methods that help us to make decisions about when to use education, marketing, or law to change public behaviors. The book concludes with major insights from the chapters, the limitations of my local research, and some thoughts about the potential for using social marketing in public emergency preparedness.

Who This Book Is For

This book is designed to be used by the emergency management community, including nonprofit and local, state and federal government emergency managers in the United States and other countries, to implement the social marketing process. Many of these professionals belong to the International Association of Emergency Managers and may also be Certified Emergency Managers. Others may be members of professional associations, such as the American Society for Public Administration's (ASPA) Section on Emergency and Crisis Management (SECM) and the International City/County Management Association's (ICMA) Center for Public Safety Management.

The book is also a resource for faculty and students in undergraduate and graduate programs in emergency management, communication, research methods, and organizational behavior or as a supplementary text in other programs, such as public or environmental health, public or nonprofit administration, and public safety management.

Because the book is a practical guide, it will complement existing texts on public behavioral change such as *Making Health Communication Programs Work: A Planner's Guide* (U.S. Dept. of Health and Human Services 2008) and *Public Awareness and Public Education for Disaster Risk Reduction: A Guide* (International Federation of the Red Cross and Red Crescent Societies 2011).

It would also complement the most popular social marketing texts including: Cheng, Kotler, and Lee's (2008) *Social Marketing for Public Health: Global Trends and Success Stories*; Lee and Kotler's (2011) *Social Marketing: Influencing Behaviors for Good*, 4th ed., Lefebvre's (2013) *Social Marketing and Social Change* and McKenzie-Mohr's (2011) *Fostering Sustainable Behavior: An Introduction to Community-Based Social Marketing*, 3rd ed. It is my hope that the book will enable emergency managers and others who attempt to change public behavior to be more successful and therefore prevent injury and loss of life.

References

American Association for the Advancement of Science Climate Science Panel. 2014. *What we know: The reality, risks, and response to climate change.*

Cheng, H., P. Kotler, and N.R. Lee. 2008. *Social marketing for public health: Global trends and success stories*. Sudbury, MA: Jones and Bartlett.

Gillis, J. 2014. Panel's warning on climate change: Worst is yet to come. *New York Times*, March 30, 2014 online. http://www.nytimes.com/2014/04/01/science/earth/climate.html.

International Federation of Red Cross and Red Crescent Societies. 2011. *Public awareness and public education for disaster risk reduction: A guide.*

Lee, N.R. and P. Kotler. 2011. *Social marketing: Influencing behaviors for good*. 4th. ed. Thousand Oaks, CA: Sage.

Lefebvre, R.C. 2012. Transformative social marketing: Co-creating the social marketing discipline and brand. *Journal of Social Marketing*, 2: 118–129.

———. 2013. *Social marketing and social change: Strategies and tools for health, well-being, and the environment*. San Francisco, CA: Jossey-Bass.

McKenzie-Mohr, D. 2011. *Fostering sustainable behavior: An introduction to community-based social marketing*. 3rd ed. Gabriola, BC: New Society.

Paton, D. and J. McClure. 2013. *Preparing for disaster: Building household and community capacity*. Springfield, IL: Charles C. Thomas.

Rothschild, M.L. 1999. Carrots, sticks, and promises: A conceptual framework for the management of public health and social issue behaviors. *Journal of Marketing*, 63: 24–37.

U.S. Department of Health and Human Services. 2008. *Making health communication programs work: A planner's guide.*

Veil, S.R., R.S. Littlefield and K.E. Rowan. 2009. Dissemination as success: Local emergency management communication practices. *Public Relations Review*, 35: 449–451.

1

THE ROLE OF PREPAREDNESS IN BUILDING A RESILIENT NATION

> What I've always found is that people will respond to meet a need in a crisis if they know what to do. You give people the opportunity to be part of something that will make a difference, and they will step up.
>
> *James Lee Witt, FEMA Director 1993–2000 (Ripley 2009, 211)*

While few people would deny that public emergency preparedness is good and important, a large percentage of the U.S. population remains unprepared. This chapter begins with a review of a recent report by the National Academy of Sciences recommending specific actions to increase disaster resilience in all sectors given our current situation (e.g., increased frequency and severity of natural emergencies, climate change, aging infrastructure, demographic changes, etc.). The next section is a brief summary of several recent national surveys on public preparedness. Some of the reasons for lack of preparedness are then discussed along with ways to overcome them.

Disaster Resilience—A National Imperative

In 2011, the U.S. government, including eight agencies and one community resilience group affiliated with a National Laboratory, asked the National Research Council (NRC) to study how to increase the nation's resilience to disasters. The NRC formed the Committee on Increasing National Resilience to Hazards and Disasters (Committee). This group of experts was asked to define "National Resilience" and how to increase it in the United States; provide goals, baselines, or metrics for resilience; describe current knowledge about resilience to hazards and disasters; and identify other information and/or obstacles necessary to increase National Resilience to disasters (National Academy of Sciences 2012, 1–2).

The Committee convened three open meetings in New Orleans and the Mississippi Gulf Coast; Cedar Rapids and Iowa City, Iowa; and Southern California. These parts of the country have all built resilience in disaster preparedness, response, and recovery (National Academy of Sciences 2012, viii).

In their final report, *Disaster Resilience: A National Imperative* (National Academy of Sciences 2012), the Committee defined National Resilience as "the ability to prepare and plan for, absorb, recover from, and more successfully adapt to adverse events" (1). They chose to fulfill their charge by preparing a vision identifying the characteristics of a resilient nation in 2030. Those characteristics were:

> In 2030, the nation, from individuals to the highest levels of government, has embraced a "culture of resilience." Information on risks and vulnerability to individuals and communities is transparent and easily accessible to all. Proactive investments and policy decisions including those for preparedness, mitigation, response, and recovery have reduced the loss of lives, costs, and socioeconomic impacts of disasters. Community coalitions are widely organized, recognized, and supported to provide essential services before and after disasters occur. Recovery after disasters is rapid and includes funding from private capital. The per capita federal cost of responding to disasters has been declining for a decade.
>
> (National Academy of Sciences 2012, 14)

Following their report, during a workshop launching a national conversation about disaster resilience (National Academy of Sciences 2013), the Committee noted the urgency of breaking from the status quo because changing conditions are causing disasters to occur more often. More people are moving into areas that are prone to natural disasters. The number of people in the United States continues to grow and includes more seniors. The nation's infrastructure is outdated. The amount of wetlands and other natural protections are declining. In part due to climate change, extreme weather events are increasing and their financial burden has outpaced the government's ability to pay for mitigation and recovery (1–2).

The report concluded with six broad recommendations to begin implementing disaster resilience at the local, state, and federal levels. While the Committee suggested who might take the lead in coordinating community resilience efforts, including communication and collaboration networks between government agencies, the private sector, and nonprofits, that crucial element remained undefined. The authority and responsibility for this effort would be a massive undertaking requiring funding and intergovernmental cooperation that can sometimes be contentious and slow.

The first of the six recommendations was for the federal government to "incorporate national resilience as a guiding principle to inform the mission and actions of the federal government and the programs it supports at all levels" (209).

The second recommendation had three components. The first encouraged cooperation between the public and private sectors to manage structural and nonstructural risk-reduction and "risk-spreading measures or tools" (212). This recommendation refers to coordination between community stakeholders on building codes and standards and on insurance that is based on local hazards.

The second component suggested that the Department of Homeland Security take the lead in working with other regional, professional, and private partners to develop "an essential framework (codes, standards and guidelines)" to address structural resilience (213). The third component was designed to heighten local awareness of hazard risks for all sectors and to reduce federal insurance subsidies through risk-based insurance pricing and multiyear policies. Those who adopted mitigation strategies could be rewarded with reduced premiums.

In its third recommendation, the Committee urged the National Science and Technology Council in the White House to work with other federal agencies, the private sector, and researchers to develop a national disaster database of "injuries, loss of life, property loss, and impacts on economic activity" (National Academy of Sciences 2012, 214). That information would be used to build quantitative models of risk and structural and social vulnerability. The fourth recommendation called for the Department of Homeland Security and other stakeholders to develop a National Resilience Scorecard that could be used as a baseline to monitor quantitative and qualitative changes in resilience. They suggested that it include a variety of factors, including the capability of infrastructure and business recovery; social capital; structural resilience to hazards, business, and market continuity; and the protection of people with special needs, including mobility, minority status, or health needs.

The Committee's fifth recommendation was that the Department of Homeland Security and the Department of Health and Human Services take the lead to help to create and sustain local and regional community resilience coalitions because of their ability to fund those efforts and provide technical support. They noted that this would require state and local government commitment to providing personnel to assist in completing the proposed National Resilience Scorecard (216). The final recommendation was that all federal agencies promote National Resilience in their programs and policies. In order to achieve this, the Committee proposed that agencies review their own policies and work together with others to fully coordinate the effort (217).

As of this writing, in mid-2014, we clearly have a long way to go to achieve the Committee's 2030 vision of resilience. That vision will require structural measures, sound and progressive policies, and a paradigm shift toward individual and community self-reliance. On December 31, 2013, as a guest on Claire Rubin's blog, *Recovery Diva*, John Plodinec of the Community and Regional Resilience Institute (CARRI) posted "Reflections on Resilience." In it, he recounted a story of foresters who were studying trees downed by wind, lightning, and disease. Although they were highly focused on the fallen trees, they recognized that they were more interested in what enabled the majority of trees to remain standing. Plodinec applies this lesson to community recovery and resilience, stating, "We must look to those still standing to understand resilience, not those who have been beaten down by the storm. If we are to become more resilient, we need to better understand what strengths are needed to recover from crises and how to nurture them." Unfortunately, we rarely take a strength-based approach to change. In this book I will draw on some of those methods. But first we'll review the most prominent surveys of public emergency preparedness.

Current Preparedness

By current standards in the United States (e.g., Ready.gov, American Red Cross), the public is not well prepared for emergencies. In 2006, the Council for Excellence in Government conducted a public readiness survey of 1,000 adults across the United States and 400 people in Chicago, Miami/Dade County, New York City, and San Francisco. The resulting Public Readiness Index (Council for Excellence in Government 2006) indicated that, of the 32 percent of people who had done nothing to prepare: "almost half (forty-five percent) simply have not thought about it; one-third (thirty-four percent) do not think an emergency will happen to them or their family; one-quarter think that nothing they can do would be effective; twenty-four percent do not want to think about it; twenty-one percent say that not knowing what to do is a major reason for their lack of preparedness; eighteen percent say it takes too much time; and sixteen percent say it costs too much money" (18).

In 2007, an American Public Health Association survey on preparedness for a public health crisis revealed that while "more than one-quarter (twenty-seven percent) of the public claim to be prepared... only half of those (fourteen percent) actually have a three-day supply of food, water, medication. . . And nearly half (forty-six percent) of the public have not pulled together a Disaster Supply Kit" (American Public Health Association 2007, 2).

In 2008, a national poll by the American Red Cross and the Council for Excellence in Government revealed that only 2 percent of the public had taken even one of the seven recommended Public Readiness Index actions defined as having a supply kit, a portable supply kit, a communications plan, a meeting place, having practiced or drilled, volunteered to help, or taken First Aid within the past five years (American Red Cross and the Council for Excellence in Government 2008, 2). According to a 2009 Citizen Corps National Survey, "fifty-seven percent of individuals reported having 'supplies set aside in their home to be used only in case of a disaster' . . . 'forty-four percent of individuals reported having a household emergency plan' and 'fifty percent of respondents reported familiarity with alerts and warning systems'" (Federal Emergency Management Agency 2009, 1–2).

Greenberg, Dyen, and Elliott (2013) attempted to identify the characteristics of people who *do* prepare. They embedded six questions about preparedness in a U.S. Department of Energy survey about the public's perception of different energy sources and nuclear waste management. Participants were asked whether they had a fire extinguisher in the home and knew how to use it, whether they had a family communication plan and a designated meeting place outside the home, a plan for sheltering at home for an extended period of time, and an emergency supply kit. The survey also included demographics, the respondent's geographical location, and questions about trust and values.

The authors hypothesized that even strong negative memories of other people's disasters (e.g., via media images) and personal experience with them may not be enough to prompt people to prepare. They suspected that in addition, people who prepared would hold conservative values such as self-reliance, a lack of trust in authority, resistance to changing the status quo, and would tend to be male. Their analysis supported the hypotheses. The difficulty with this group of prepared individuals lies with their potential lack of willingness to work with government-sponsored and community-preparedness groups and with others who do not hold their values. Greenberg, Dyen, and Elliott (2013) declined to suggest a role for these people, leaving that decision to local governments who can weigh the pros and cons of involving them in preparedness activities (e6). I have not seen other data that corroborates this study. It appears that these individuals are "survivalist" types, unlike emergency preparedness volunteers such as Community Emergency Response Team (CERT) members across the United States, who, I suspect, vastly outnumber the former.

In order to determine disaster preparedness levels among older adults in the United States, Al-rouson, Rubenstein, and Wallace (2014) surveyed a subset of the 2010 Health and Retirement Study (HRS). The HRS is a "nationally representative cohort study that surveys the social, economic, and health characteristics of Americans aged fifty and older" (506). The researchers analyzed the results of eighteen disaster preparedness indicators within a "preparedness module" survey given to a portion of the respondents to the larger core survey.

They found that "about two-thirds of the study population had no emergency plan, had never participated in any disaster preparedness educational program, and were not aware of the availability of relevant resources. More than one-third of the respondents did not have a basic supply of food, water, or medical supplies . . . such deficits could result in further decline in health status, especially in the presence of mobility and functional limitations" (510). They emphasize that in addition to those previously mentioned, poverty, lack of transportation and communication equipment, and low educational attainment highlight the need for disaster plans specifically for older adults.

In a comprehensive review of death records from Hurricane Katrina in Louisiana, Brunkard, Namulanda, and Ratard (2008) documented at least 971

Katrina-related deaths in Louisiana and 15 evacuee deaths. They found that 49 percent of the victims in Louisiana were people 75 years old and older, "whereas their age cohort represents fewer than six percent of both the greater New Orleans and the overall Louisiana population . . . a total of 178 people seventy-five years old and older died in their homes, 115 from drowning" (5). The authors suggest at least two explanations, both of which likely contributed to the death toll. One is that older people may not have left their homes because they had evacuated before when the intensity of the storm was not as predicted and because of fear of looting or being unwilling to disrupt their routines. A second is that comorbidities caused drowning or injury (6). Many older persons develop physical and mental health conditions that make performing basic activities difficult on a normal day, much less under emergency conditions. Another important cause of injury and death during disasters is poverty.

Ripley (2009) emphasized the terrible effects of poverty, "Of all the people who died from natural disasters on the planet from 1985 to 1999, 65 percent came from nations with incomes below $760 per capita, according to the Intergovernmental Panel on Climate Change" (McCarthy et al. 2001). She compared the 1994 Northridge earthquake in California to the 2005 earthquake in Pakistan that was similar in magnitude. The Northridge earthquake killed only 63 people, while the Pakistan earthquake killed about 100,000 (90). Ripley cites a study by Matthew Kahn (2005) at Tufts University who found that raising national GNP from $2,000 to $14,000 per person can save the lives of 530 people during a disaster. Money, which he called "liquid resilience," also helps survivors to recover more quickly (Ripley 2009, 90).

Why Don't People Prepare?

Lack of preparedness is due to a variety of factors, including not having any prior experience with an emergency or disaster, incorrect perception of actual risks, expectations that public safety forces will be able to help everyone all the time, a lack of information about how to prepare, feeling that there is nothing that can be done to prepare, and a lack of resources. Additional factors are a failure to exercise (i.e., drill or practice) plans that are in place, such as sheltering-in-place during tornado warnings. Because preparation is voluntary, other activities easily take priority in people's lives. But Paton and McClure (2013) insist that it is not enough to know that some people have not prepared. We need to discover their reasoning in order to know how to help them to take action.

It is also important to analyze the actions that people say they have taken. Paton and McClure (50) cited Prior's (2010) study of wildfire readiness, which showed that rather than preparing for a specific hazard, people are more likely to prepare with items that they use routinely, such as food. Lopes (1992) and other researchers have followed up self-reports of household preparedness and found differences between what people said they had done and actual preparations (Paton and McClure 2013, 53).

Paton and McClure (2013) discuss several reasons for "selective preparedness." First, homeowners have a greater investment in their homes than renters, so they are more likely to take action to protect their property, while renters frequently lack the resources or freedom to make structural improvements to their living quarters or to store emergency supplies. They note that Diekmann and Preisendörfer's (1992) low-cost hypothesis "proposes that people are generally more likely to adopt low-cost, relatively ineffective actions over higher-cost but more effective actions" (57). These are items such as food, flashlights, and a barbeque. Paton and Johnston (2008) addressed this conflation of everyday and hazard items by asking people how they would use the items in an emergency and also whether they had "changed their shopping habits to gradually increase their emergency food supplies" (Paton and McClure 2013, 58).

Everyday Risks versus Disasters

Local emergency managers are often frustrated in their efforts to encourage the public to take personal responsibility for their safety and survival. Simply communicating risk and providing information usually does not cause people to take action. Because of segmentation in the roles of emergency management, the fire service, public health, and other government agencies, risk is not communicated in a holistic way. Doing so might help people to understand their risk for a car accident or heart attack due to their lifestyle and, for example, potential injury or death from a flood or tornado due to their geographic location. However, progress is being made. Currently, some health professionals ask their patients (e.g., socially isolated senior citizens) if they have an emergency plan.

As Paton and McClure (2013) note, "people's concern about risks often bears no relationship to the objective probability of being harmed by those risks (Slovic, Fischhoff, and Lichtenstein 1982)" (42). In part, the media causes these misperceptions because it exposes people to events that are less common than falls and strokes, and so they become more fearful of the former than what is most likely to happen. Tversky and Kahneman (1982) defined this human tendency as the availability heuristic.

According to Ripley (2009), "Usually people think . . . either something will happen to me or it won't. . . . So when people hear that they have a 6 in 100,000 chance of dying from a fall, they shelve that risk under the label 'won't happen to me,' even though falling is in fact the third most common cause of accidental deaths in the United States (after car crashes and poisonings)" (46). Ripley's data was taken from *Deaths: Final Data for 2003, National Vital Statistics Reports*, 54, no. 3 (April 19, 2006). The most current data is available from the National Vital Statistics website at http://www.cdc.gov/nchs/products/nvsr.htm.

Preparedness depends on hazards, the people involved, the situation, infrastructure, and the "hazard-scape"—the hazard characteristics and behaviors that can change over time with new information (Paton and McClure 2013, 6). The U.S. population densities are shifting to the southern and coastal regions of the country, putting more people at risk for weather emergencies. Experiencing emergencies and disasters changes how people react to warnings in a range of ways, from more active preparation to ignoring messages altogether when they have been asked to evacuate one too many times.

In the United States, "fire is the single most common disaster nationwide. The American Red Cross responded to more than 70,000 disasters last year [2013]; ninety-three percent were fire-related" (February 2014 ARC Emergency News RC-34–0220). Ripley (2009) also stresses that fire and poverty are linked. She talked with Denis Oniel, who began his career as a firefighter in Jersey City, New Jersey, in 1971 and is now superintendent of the National Fire Academy. Oniel told Ripley that he "never fought a fire in a rich person's home" noting that fires are more common in poor neighborhoods where people live in substandard housing without smoke detectors and use portable heaters for warmth (90).

Tragically, Ripley (2009) reported that 25 percent of the people who die in fires are African American, "twice their share of the population" and "African American and American Indian children are nearly twice as likely to die in a fire than white or Asian children" (89). As of this writing, the most recent fire statistics from 2003–2007 (Federal Emergency Management Agency, U.S. Fire Administration—National Fire Data Center 2009) indicate that the "relative risk of children age 4 and younger dying in a fire is slightly less than that of the general population. In the future, additional analysis is required to confirm that this is a true trend change" (5).

The report also states that "African-Americans and American Indian males have much higher fire death rates than the national average" (4). In terms of gender, "approximately fifty percent more men die in fires than women" (5). Possible reasons for this are that men are more likely to be intoxicated and also work at more dangerous professions. Nearly one-third of women's fire deaths are in the seventy years old and older group. "Finally, people with limited physical and cognitive abilities, especially older adults, are at a higher risk of death from fire than other groups" (Federal Emergency Management Agency, U.S. Fire Administration—National Fire Data Center 2009, 5).

Although the report emphasizes that fires continue to decline in the United States due to an increase in smoke alarms and sprinklers, adherence to fire codes, more fire proof construction materials, public education, and better firefighting equipment, the fire death rate in the United States is "two to two and one half times that of several European nations and at least twenty percent higher than many other Nations. Of the twenty-five industrial Nations examined by the

World Fire Statistics Centre, the United States ranked as having the fifth highest fire death rate. This general status has been unchanged for the past twenty-seven years" (2). This alarming situation is my rationale for including fire prevention as a component of public education for emergency preparedness.

According to the National Weather Service's *Summary of Natural Hazard Statistics for 2012 in the United States* (2012), the ten-year average (2002–2011) number of weather-related deaths is 641 (3). But this can vary dramatically from year to year depending on the type and number of weather events and temperatures. And as with deaths by fire, almost twice as many men die than women. According to the report, this is likely because of the greater "percentage of men who hold outdoor jobs such as construction and who take part in sports and other outside activities such as fishing and boating. In 2012, men of all ages were more likely to be victims except in the ninety plus age category," in which there are more women (3).

The 2014 update of the EPA's ongoing report, *Climate Changes Indicators in the United States*, includes thirty main indicators divided into six categories: greenhouse gases, weather and climate, oceans, snow and ice, health and safety, and ecosystems. The last two categories include four new indicators, heating and cooling days, Lyme disease, wildfires, and water levels and temperatures (Homeland Security Digital Library Blog 2014). The increase in climate change indicators highlights the importance of public preparedness and awareness.

There are also nuances within weather emergencies that make some situations more deadly. For example, according to Ready.gov/floods, "floods are one of the most common hazards in the United States, however not all floods are alike. Some floods develop slowly, while others such as flash floods can develop in just a few minutes and without visible signs of rain. Additionally, floods can be local, impacting a neighborhood or community, or very large, affecting entire river basins and multiple states" (www.ready.gov/floods).

Paton and McClure (2013) note that "in general, people take more precautions for high-frequency events [e.g., theft], and take fewer precautions for low-frequency events, even if the low-frequency events carry higher potential losses (Slovic, Fischhoff, and Lichtenstein 1982)" (37). Faulty perception of risk highlights the importance of risk communication. Paton and McClure (2013) stress that "unless people hold a belief that a hazard event will occur in the immediate future (e.g., a few months to a year), it is unlikely that they will take steps to mitigate their risk, particularly as they may have other, more immediate and more salient pressures in their daily lives" (39). They recommend that people be told the benefits of preparing before they are given a list of things to do so that the cost of preparation is not the foremost thing in their minds.

Paton and McClure cited Hassol (2008) who "showed that several terms used by scientists to describe risks and hazards have a different meaning for the public" (41). For example, the public may interpret the term "uncertainty" as scientists not knowing if they are correct and "error" as if something is incorrect, rather than the higher and lower limits of an event. They advise specifying limits directly and if there is a positive trend in something to say upward, as with temperatures under climate change, because the public may interpret positive as good. They add that incorrect perception of risk can be countered by saying that there is converging scientific "evidence from a number of sources" (41) and to use both words and numbers to convey information because that is further clarification (42).

Public emergency preparedness that fosters resilience requires more than the basics of having a kit, making a communication plan, and knowing where to get information. The fourth activity noted on the Ready.gov website, getting involved, enables people to practice skills that will be necessary in an emergency. Ripley (2009) and others have emphasized that the best preparation is through "repeated, realistic training" (75). She believes that children tend to get the best training for disasters because of actual practice, for example, in fire drills and "stop, drop, and roll." These responses become imbedded in the subconscious and this is critical to survival. Ripley quotes Richard Gist, a psychologist with the Kansas City Fire Department, "If you have to stop and think it through, then you will not have time to survive" (126).

Paton and McClure (2013) emphasize that three readiness strategies enable people to respond rather than simply react in emergencies. These strategies focus on developing people's ability to a) anticipate hazard consequences, b) cope and recover from those consequences, and c) to learn from their experiences (46).

One way for the public to prepare their family and community for emergencies is by volunteering with their local Community Emergency Response Team (CERT). CERT is one of several programs under the federal volunteer coordination agency Citizen Corps that is administered by the Federal Emergency Management Agency (FEMA). CERT was founded by the Los Angeles Fire Department (LAFD) in 1985. The LAFD recognized that people will spontaneously volunteer in disasters, but if they are not trained, they may become casualties. For example, during the 1985 earthquake in Mexico City, although civilians rescued at least 800 people, over 100 died in the effort. CERT was later adopted by FEMA. The CERT curriculum is standardized so that teams across the nation all receive the same initial training.

The purpose of CERT training is to provide volunteers with the basic skills they will need to respond to their community's needs before, during, and after a disaster when emergency services may not be available. CERT basic training is twenty hours long and topics include emergency preparedness in the home and workplace, fire safety, medical operations, light search and rescue, disaster psychology, CERT organization, CERT and terrorism, and a final disaster simulation exercise. There are over 2,200 registered CERT programs in the United States as of this writing (www.fema.gov/community-emergency-response-teams).

CERTs typically fall under the jurisdiction of their local fire chief or emergency management agency director who may deploy the CERT during emergencies. Frequently, communities use a CERT's capabilities for nonemergency

FIGURE 1.1 CERT volunteers in action (Image courtesy of William C. Rieter)

events such as directing traffic, distributing preparedness information, or assisting with fire drills and other exercises. Although not all first responders are comfortable involving citizens during emergencies, the CERT program has been so successful because many recognize the value of extra trained personnel during emergencies. The use of volunteers for public emergency preparedness will be discussed further in later chapters.

Conclusion

This chapter provided a macro perspective on public preparedness in the United States as a foundation before we delve into the challenges of individual and social change using social marketing. Strong relationships within communities, from neighborhoods to the federal government, are a key component of disaster resilience, as highlighted by the National Academy of Sciences report discussed earlier. Such relationships enable good communication and support before, during, and after emergencies.

It is the relationships formed between people during the social marketing process that make it so much more successful in achieving lasting change than one-way, one-time communication or training. Of course, getting to know and collaborate with people takes time. But in terms of personal and public preparedness, such relationship building enhances understanding and trust, which are vital during crisis conditions. In the next chapter, we begin our journey to readiness.

References

Al-rousan, T.M., L.M. Rubenstein, and R.B. Wallace. 2014. Preparedness for natural disasters among older U.S. adults: A nationwide survey. *American Journal of Public Health*, 104(3) (March): 506–511.

American Public Health Association. 2007. *National opinion survey to determine levels of preparedness for a public health crisis*. http://www.safezonellc.com/documents/Survey_Report.pdf.

American Red Cross and the Council for Excellence in Government. 2008. *Readiness Quotient National Poll*. http://www.volunteerwv.org/nd/assets/downloads/CC/Readiness_Survey_Results_July08.pdf.

Brunkard, J., G. Namulanda, and R. Ratard. 2008. Hurricane Katrina Deaths, Louisiana, 2005. *Disaster Medicine and Public Health Preparedness*. Published online August 28, 2008. http://www.ncbi.nlm.nih.gov/pubmed/18756175/.

Council for Excellence in Government. 2006. Are we ready? Introducing the Public Readiness Index: A survey-based tool to measure the preparedness of individuals, families and communities. http://www.fema.gov/media-library/assets/documents/31690.

Diekmann, A. and P. Preisendörfer. 1992. Personliches Umweltverhalten: Diskrepanzen Zwischen Anspruch und Wirklichkeit. *Kölner Zeitschrift Für Soziologie und Sozialpsychologie*, 44: 226–251.

Federal Emergency Management Agency. 2009. *Personal preparedness in America: Findings from the 2009 Citizen Corps National Survey, Summary Sheet*. (Revised December, 2009).

Federal Emergency Management Agency, U.S. Fire Administration—National Fire Data Center. 2009. *Fire in the United States 2003–2007*. 15th ed. FA-325/October.

Greenberg, M.R., S. Dyen, and S. Elliott. 2013. The public's preparedness: Self-reliance, flashbulb memories, and conservative values. *American Journal of Public Health*, 103(6): 85–91.

Hassol, S.J. 2008. Improving how scientists communicate about climate change. *Eos*, 89(11): 106–107.

Homeland Security Digital Library Blog. 2014. *EPA adds new indicators in updated climate change report*. Posted online May 30, 2014. https://www.hsdl.org/blog/newpost/view/epa-adds-new-indicators-in-updated-climate-change-report.

Kahn, M.E. 2005. The death toll from natural disasters: The role of income, geography and institutions. *Review of Economics and Statistics*, 87(2): 271–284.

Lopes, R. 1992. *Public perception of disaster preparedness presentations using disaster images*. Washington, DC: The American Red Cross.

McCarthy, J.J., O.F. Canziani, N.A. Leary, D.J. Dokken, and K.S. White. 2001. *Climate change 2001: Impacts, adaptation, and vulnerability*. Contribution of Working Group II to the Third Assessment Report of the Intergovernmental Panel on Climate Change: Insurance and other financial services, 451–486. Cambridge, UK: Cambridge University.

National Academy of Sciences. 2012. *Disaster resilience: A national imperative*. Washington, DC: National Academies.

———. 2013. *Launching a national conversation on disaster resilience in America: Workshop summary*. Washington, DC: National Academies.

National Weather Service. 2012. Summary of natural hazard statistics for 2012 in the United States. www.nws.noaa.gov March 12, 2014.

Paton, D. and D. Johnston. 2008. *A means-end chain theory analysis of hazard cognitions and preparedness*. Wellington, New Zealand: Earthquake Commission.

Paton, D. and J. McClure. 2013. *Preparing for disaster: Building household and community capacity*. Springfield, IL: Charles C. Thomas.

Prior, T. 2010. *Householder bushfire preparation: Decision making and the implications of risk communication*. Unpublished Ph.D. dissertation, University of Tasmania, Hobart, Australia.

Ripley, A. 2009. *The unthinkable: Who survives when disaster strikes—and why*. New York: Three Rivers Press.

Slovic, P., B. Fischhoff, and S. Lichtenstein. 1982. Facts versus fears: Understanding perceived risk. In *Judgment under uncertainty: Heuristics and biases*. D. Kahneman, P. Slovic, and A. Tversky, eds., 463–492. Cambridge: Cambridge University.

Tversky, A. and D. Kahneman. 1982. Judgments under uncertainty: Heuristics and biases. In *Judgment under uncertainty: Heuristics and biases*. D. Kahneman, P. Slovic, and A. Tversky, eds., 3–20. Cambridge: Cambridge University.

2

THE CHALLENGE

Moving from Complacency to Readiness

> Among the most important evidence-based lessons regarding risk communication are three key ideas: (1) recruit prepared members of the public to be advocates, (2) tell people what they should do, and (3) repeat the message.
>
> *(Drabek 2013, 272)*

This chapter provides an overview of the current behavioral change models used for emergency preparedness communication and training in the United States, along with models used in some other countries. That overview is followed by an explanation of social marketing and a few instances when it has been used in emergency preparedness.

The Citizen Corps Personal Behavior Change Model

Communicating risk and informing people how to prepare for disasters is an important first step in protecting the public, but it is quite another to convince people to take specific actions. The Citizen Corps Personal Behavior Change Model for Disaster Preparedness (PDP model) "identifies segments of the population based on their perceptions of threat and efficacy and provides associated areas of focus for outreach and social marketing that target specific barriers and motivations. It was developed by Macro International, Inc. under contract with FEMA's Community Preparedness Division." The PDP model is based on . . . the 1) Extended Parallel Process Model (EPPM) [Witte 1998] and 2) the Stages of Change/Transtheoretical Model" [Prochaska and DiClemente 1982] (Federal Emergency Management Agency 2006, 2).

Although the definition of the PDP model mentions social marketing, the model was a first iteration and was to be validated through surveys and focus groups. It was "intended to serve as a tool to help design successful outreach/social marketing approaches and as a framework to conduct further research into the motivating factors and barriers to personal preparedness" (1). The EPPM focuses on individual risk perception and response. If people do not perceive a risk, they will not take action. If they do perceive a risk, they will only take appropriate action if they believe they can be effective. If not, they will avoid taking action by denying or minimizing the risk. Depending on the hazard, an individual's response may fall into one of four categories: "a) person does not believe the he or she is susceptible to the threat ... b) person does not believe that he or she is presented with a severe threat ... c) person either does not know what the recommended actions are or does not believe that he or she can perform the recommended protective actions ... d) person does not believe that the recommended protective actions will be effective" (3).

Within the PDP model, EPPM can be used to divide people who are not motivated to prepare into one of three "Threat/Efficacy Profiles" (Federal Emergency Management Agency 2006, 3). The three profiles identified those who thought they would not be affected and so were unprepared, those who understood their danger but had barriers to preparation and were unprepared, and those who understood the danger and felt capable of preparing. This population segmentation facilitates the development of targeted messages and outreach to each group to enable them to take a next step toward preparedness.

The Stages of Change/Transtheoretical Model (TTM) includes five stages of readiness to take action: precontemplation, contemplation, preparation, action, and maintenance (4). As with the EPPM, the TTM is used to group individuals in a specific stage of change in order to take actions that will move them to the next stage of readiness. Precontemplation includes those who are not thinking about preparing or have no intention of doing so (typically within the upcoming six months). Those who are contemplating action, plan to do so within the next six months. If people are in the preparation stage, they are "actively considering changing [their] behavior in the immediate future," typically within a month (4). People in the action stage "have actively made an overt behavioral change ... but the changes are not well established (maintained for six months or less)" (Federal Emergency Management Agency 2006, 4). Finally, those in the maintenance stage have continued their behavior for more than six months and intend to do so in the future. Successful application of the PDP model should lead to individual emergency preparedness actions that are maintained over time that will foster a "true and sustained culture of preparedness" (Federal Emergency Management Agency 2006, 12).

Communication scholars Paek et al. (2010) conducted a statewide citizen emergency preparedness survey of 1,302 adults in Georgia using the TTM and noted that "progression from even one stage to the next is a step in the right direction, but ... efforts should be aimed at not just at ushering people from contemplation to action, but from action to maintenance" (440).

In their work on disaster preparedness, Paton and McClure (2013) identified four theories of behavior change and three theories of change specific to natural hazards. The four behavior change theories were the TTM discussed earlier and the health belief model (HBM), theory of planned behavior (TPB), and protection motivation theory (PMT). The three theories of change were the Person-relative-to-Event (PrE) model, the Critical Awareness Model, and the Protective Action Decision Model (PADM). They emphasized that these theories "suggest that effective action results from providing people with both high quality information about hazard(s) and their mitigation *and* [italics in original] ensuring they have the personal and social competencies required to interpret and use the information to guide their adoption decisions" (97).

The HBM is defined in *Making Health Communication Programs Work* (U.S. Department of Health and Human Services 2008) as "a conceptual framework of health behavior stating that health behavior is a function of both knowledge and motivation. Specifically, the model emphasizes the role of perceived vulnerability to a condition, perceived severity of the condition, perceived benefits of the recommended action, perceived barriers to the advised action, cues to action, and self-efficacy in terms of one's ability to take action" (248). Paton and McClure (2013) reported that Dooley, Catalano, Mishra, and Serxner (1992) used the HBM to analyze earthquake preparedness and found that "the relative contribution of barriers versus benefits influenced what people did" (99).

Paton and McClure employed Azjen's 1991 theory of planned behavior (TPB), which "proposes that behavior is a product of intentions that are predicted in turn by three factors: people's attitude toward the target behavior, their "perceived subjective norm" (which includes their judgments about social pressures to perform an action), and their perception of behavioral control (and in some later versions self-efficacy), which refers to people's perception of how difficult it is to perform the target behavior" (2013, 105).

Protection motivation theory (PMT) "results from the interaction between two processes: one concerned with threat appraisal, and the other with coping appraisal" (99). Threat appraisal includes perception of the hazard's severity and one's vulnerability to its consequences. Coping appraisal is the interaction between beliefs about protective response efficacy and self-efficacy. People weigh the costs against the benefits, and then make decisions about taking action. Grothman and Reusswig (2006) used PMT to predict flood preparation behavior. Paton and McClure interpreted their findings as suggesting that information should be provided about risks, but also about the costs of preparation and especially why it would be beneficial so that people have positive information when making cost-benefit decisions (101).

The Person-relative-to-Event model (PrE) "argues that problem-focused coping influences behavior if the magnitude of a threat is exceeded by the person's resources to deal with the threat. If people believe that their resources are high, then when the threat increases preparedness increases" (103). The opposite is also true.

Paton and McClure report that Mulilis and Lippa's 1990 research on earthquakes supports this model (Paton and McClure 2013, 103).

Paton, Smith, and Johnston (2005) developed the Critical Awareness Model after studying how people adapted to the effects of volcanic eruptions. Their model is based on Dalton, Elias, and Wandersman's (2001) theory of critical awareness, which "describes the extent to which people think and talk about a specific source of adversity or hazard within their environment" (Paton and McClure, 103). Paton, Smith, and Johnston (2005) found that hazard preparedness depends on the importance it has for people, but it is also socially constructed. Their theory "proposed that the relationship between motivation to act and intentions to prepare was mediated by outcome expectancy (also called response efficacy), self-efficacy, and action-coping variables . . . and personal capabilities and resources, . . . trust, willingness to accept personal responsibility for their safety . . . and beliefs regarding when the next hazard event would occur" (Paton and McClure 2013, 104–105).

The Protective Action Decision Model (PADM) is similar to the Critical Awareness Model and "argues that people's decisions about how to respond . . . depend on judgments they make about three factors; the threat posed by the hazard, identifying the actions available and capable of protecting them against the hazard consequences, and identifying the sources from whom information about hazard and protective measures can be obtained" (111). The PADM is detailed because it accounts for uncertainty both before and during the decision-making process. People may not seek information if they are unaware of a hazard due to their social isolation or poor information from their peers. They may not know where to get information once they decide to, or it may be from an untrusted source so they ignore it. Once a decision is made to take preparedness actions, they may not have the skills, resources, or sense of self-efficacy to effectively complete them. For a detailed discussion of this model, see Paton and McClure, 111–116.

Finally, although Paton and McClure note that "social marketing is increasingly being used to guide the development of hazard public education programs" (109), their discussion of the process is brief. They identify it and the other seven models and theories as social-cognitive theories in Figure 5.1, (Paton and McClure 2013, 95). With this foundation in behavioral change models for disaster preparedness we turn to an in-depth exploration of social marketing.

Social Marketing—Societal versus Individual Behavioral Change

Lefebvre (2013) notes that although many social marketing efforts focus on individual behavior, broader social change requires additional processes. The most common models used in social marketing are two that were discussed previously, the stages of change and health belief models. In addition, Lefebvre suggests the use of social-cognitive theory (Bandura 1986) and diffusion of innovations theory (Rogers 1995) (79). He outlines both theories briefly in Table 3.1 (Rogers 1995, 80–82).

From that table, social-cognitive theory includes reciprocal determinism (bidirectional interaction between a person and the environment), behavioral capacity (knowledge and skills), expectations (beliefs about the results of actions), self-efficacy (confidence to take action), and observational learning and reinforcement (responses to increase or decrease in behavior). Diffusion of innovations theory offers relative advantage (over current behaviors or products, etc.), is compatible with existing routines and values, is not complex (easy to do), gives people a chance to try the new behavior before they adopt it, and is easily identified and observable by others. Lefebvre refers to the wicked social problems that we are trying to address as "puzzles" and advocates for using the theories or combinations of theories that best suit the people and situations before us.

Rather than encouraging someone to change, which inevitably causes resistance, Lefebvre suggests asking "Can I help you learn something new that might help you solve a problem better than what you are doing now?" (84). Taking this approach, we can then rely on proven learning theory to focus our efforts. Lefebvre explains Bandura's (1986) six-step model of how we usually learn as first seeing or experiencing the behavior we want to attempt. Because of our interest, we pay attention to that behavior and then remember it. The fourth and fifth steps are believing in our ability to do it and wanting to do it. The final step is attempting the behavior (Lefebvre 2013, 84). However, this is a model for individual change, and we are seeking to change the behavior of populations (e.g., public preparedness in a neighborhood). Lefebvre emphasizes that we need to think much more broadly about social change. Individuals are important, but we also must include "social networks, formal and informal organizations, markets, and government regulations and policies" (93).

Lefebvre and Kotler (2011) applied concepts from behavioral economics to social marketing. The behavioral economics literature (e.g., Thaler and Sunstein, 2008 and Kahneman, 2011) counters mainstream economics, which assumes that people make rational decisions, weighing the pros and cons. Behavioral economists recognize that many, if not most, human decision making is based on unconscious emotions and imitating those around us. One of Lefebvre and Kotler's (2011) lessons for social marketers is to remember that people have a variety of reasons for changing their behavior, so they need to be approached in a variety of ways, for example, through the traditional method of the 4 Ps: Product, Price, Place, and Promotion. Another lesson is to recognize that the way a problem is framed, as in political communication, will affect how it is received and acted upon. Their final lesson is that markets are not efficient or rational; hence, there are inequities in health and social justice (Lefebvre 2013, 94). Before we go further into the social marketing process, a brief history of the field will provide some context.

Origins of Social Marketing

The term social marketing has been confused with the use of social media and social networking. Cheng, Kotler, and Lee (2008) state that social media and networking may be used as tools in social marketing, but they are not true social marketing as it has been practiced since its founding (11). According to Lee and Kotler (2011), the term *social marketing* was first introduced by Philip Kotler and Gerald Zaltman (1971) to describe "the use of marketing principles and techniques to advance a social cause, idea or behavior" (2011, 12). Andreasen (1995) defined social marketing as "the application of commercial marketing technologies to the analysis, planning, execution, and evaluation of programs designed to influence the voluntary behavior of target audiences in order to improve their personal welfare and that of their society" (7). Alternatively, Lefebvre (2013) emphasizes that rather than being a behavioral theory, social marketing is a "systematic approach to thinking about and solving the wicked problems our world faces" (5). These wicked problems are often based on societal inequalities and other factors that individuals cannot change on their own.

Andreasen (2002) notes that the field is now widely accepted with several general textbooks, chapters in basic marketing textbooks, its own journal, three annual conferences, and social marketing centers in several parts of the world, along with training programs (3). Social marketing has been used by U.S. federal agencies, UNAIDS, and by nonprofit organizations and private consulting firms. Cheng, Kotler and Lee (2008) state that "social marketing has been widely used in solving public health problems, has fast become 'part of the public health domain' (Ling, Franklin, Lindsteadt, and Gearon 1992, 360) and will 'play a bigger role in public health' (358)" (4–5). Two examples of social marketing's prominence in public health are the Centers for Disease Control and Prevention (2013) website *Gateway to Health Communication & Social Marketing Practice* and *Making Health Communication Programs Work: A Planner's Guide* (U.S. Department of Health and Human Services 2008).

The Social Marketing Process

Nancy Lee and Philip Kotler (2011), two of the leading social marketing researchers, emphasize that there is a strong temptation to forego taking the time to determine whom you want to reach, which specific behaviors you want to help them to change, and how to do that by addressing their needs and concerns. Fortunately there are many versions of the social marketing planning process on the web but, with their permission and that of their publisher, Sage, I have chosen to use the following ten steps, which are taken from Appendix A (459–471) in *Social Marketing: Influencing Behaviors for the Good* (2011). I will discuss the process as a first step toward preparing a full social marketing plan for public emergency preparedness later in the book. The setting is my local county in the Midwestern United States, and we will focus on the first recommended preparedness step on Ready.gov: build a seventy-two-hour kit. Ready.gov advises having both a home and go kit, but this plan will focus solely on the home kit.

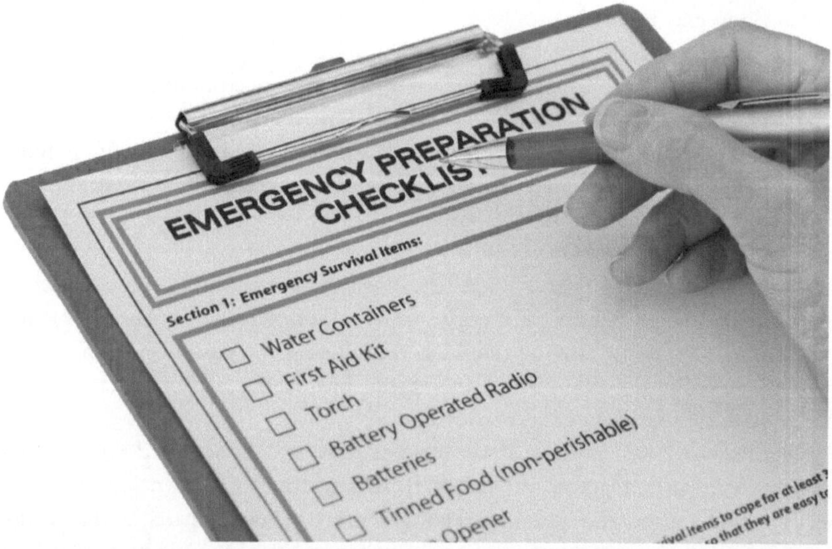

FIGURE 2.1 Using a simple checklist helps reduce resistance to preparation (Image courtesy of Pixooz/iStock)

Lee and Kotler (2011) note that their social marketing plan differs from mainstream private sector marketing in three ways. The first is that the target audiences, which I will, following Lefebvre (2013, 53), call the priority audiences, are selected before the behavioral goals are established. This seems odd, but think of the difference between how a mother with young children would prepare versus an elderly man who lives alone. A second difference is that "competition" with the desired behavior is not identified at the outset but rather about midway through the process. And a third difference is that the purpose of the plan, in this case, public preparedness by having a seventy-two-hour kit at home, is not the goal. Rather it is the number of people in the priority group who actually adopt the behavior (have a kit).

Step 1: Describe the Background, Purpose, and Focus

The social issue that this plan is addressing is lack of public emergency preparedness, specifically not having a seventy-two-hour emergency kit at home. Because our county has not been surveyed on preparedness, we are assuming that lack of preparedness is comparable to the national average or less because this area has not experienced major disasters such as hurricanes or earthquakes. The most common emergencies are flooding, snowstorms, and severe rainstorms.

However, with the onset of climate change, we know that some effects may be extremes in weather patterns, wildfires, changes in ground and surface water levels, movement of wildlife and flora, disruption of the growing season, and an increase in natural disasters. These factors, along with others noted by the National Academy of Sciences report, *Disaster Resilience: A National Imperative* (2012) discussed previously, warrant the public becoming a partner in emergency preparedness and National Resilience.

The benefit of creating and maintaining a seventy-two-hour kit is the ability to survive for at least three days when an emergency occurs and first responders may be overwhelmed. Having a kit also enables individuals and families to help others during a crisis because their basic needs are met.

Step 2: Conduct a Situation Analysis

Because this is a planning exercise, Lee and Kotler suggest conducting a SWOT (strengths, weaknesses, opportunities, and threats) analysis (2011, 43). In our case, we identified strengths, such as our twenty-seven CERT teams, who had already prepared for emergencies and our relatively strong philanthropic sector as a source of resources. Weaknesses included some groups that we anticipated were the least prepared, for example, the poor, people with functional needs, and socially isolated seniors. Another weakness was a lack of staff and resources to carry out the social marketing planning process and the execution of the plan itself.

Our Local Emergency Planning Committee (LEPC) and county Office of Emergency Management (OEM) were both sources of opportunities, the former due to their prior activity in training the public and the later due to their enthusiasm for the project and expertise. The largest threat was attempting to launch a new public preparedness behavioral change process with little documented success, although it had been proven to be effective in public health and other areas. Lee and Kotler note that important threats include "cultural, technological, natural, demographic, economic, political, and legal forces" (43). They suggest that it is valuable to spend time researching what others have done as part of the SWOT analysis.

Step 3: Select Target Audiences

Selecting the target, or priority, audiences is a very important component of the planning process and consists of three steps. The first is segmenting the population, in our case, the public of our county, into smaller similar groups. The segments are then prioritized based on criteria related to the mission. In our case, we used Lefebvre's (2013) criteria: "1) Who are the people at highest risk? 2) Who are the people most open to change and 3) Who are the critical-for-success groups?" (129). Initially, we believed that the people at highest risk were the groups previously mentioned: the poor, people with functional needs, and socially isolated seniors, and also subsets of those groups. We identified these groups based in part on the literature about public preparedness but also from our prior work with the county LEPC, OEM, first responders, and Red Cross, and from our county demographics.

Step 4: Set Behavior Objectives and Goals

In this step, three types of goals and objectives are outlined: those for behaviors, knowledge, and beliefs. We may be asking people to "accept . . . reject . . . modify or abandon. . . ." (44) some behaviors. In our case, we are planning to help people to accept the behavior of making and maintaining a seventy-two-hour home kit but also to abandon their total reliance on first responders. The knowledge that they will gain is an appreciation for the risks and hazards in our community and how to make and maintain a kit. The primary belief objective is that their preparedness efforts are beneficial during an emergency because first responders can focus their efforts on those most in need. Lee and Kotler suggest the use of SMART (specific, measurable, attainable, relevant, and time sensitive) goals during this process (44). We will define these more extensively in later chapters.

Step 5: Identify Target Audience Barriers, Benefits, the Competition, and Influential Others

During this step, you begin to build a collaborative learning relationship with your priority group(s). They are your most important resource for identifying barriers to preparation, the benefits of doing so, the activities, and priorities that compete with preparation and influential others, such as family members, friends, and perhaps their children. With this information you can design messages and training that meet the specific needs of those you wish to help.

Step 6: Develop a Positioning Statement

The positioning statement builds on the information you gathered in step five to describe how you want the priority group to think about the new behaviors compared to the competing actions. Most people are familiar with the marketing term "branding." For our purposes, an example of a position statement would be: We want seniors to know how to prepare a seventy-two-hour home kit and to realize the benefits so that they make and maintain one.

Step 7: Develop a Strategic Marketing Mix (4Ps)

In this step we move into the marketing planning phase by identifying the 4Ps: Product, Price, Place, and Promotion. This is an important and detailed step. Lee and Kotler (2011) also characterize the marketing mix as "the determinants (independent variables) used to influence behaviors (the dependent variable)" (46). They stress the importance of developing the mix in the sequence outlined earlier, so we'll begin with the product, which is subdivided into core, actual, and augmented. In our case, the core product includes the benefits of building a kit that the priority group has identified as valuable. The actual product is making and maintaining the kit itself. The augmented product would be any other objects or services in addition to the kit, for example, opportunities to participate in additional training.

The price, of course, would be the cost of the materials to build a kit. If materials are donated, those would be offered as incentives. There are currently no monetary disincentives, such as fines, for not having a kit, nor any incentives, such as a discount on insurance premiums. Place refers to the site where the action will be taken and also, in some cases, when it is done. For example, some people might find it easier to assemble a kit in a group with friends, such as at a community center, if the components were provided, while others may do fine on their own at home. Finally, Lee and Kotler (2011) define promotion as the messages that you will use to communicate, the people who will deliver those messages (e.g., the agency, partners, elected officials), and how that information will be delivered (48).

We would focus on key messages for the priority groups defined earlier, crafting the message using their own words and thoughts about the benefits of preparation. The messengers would be the leaders in the organizations that our priority groups are members of, our community partners, and others who we have discovered are influential within the priority groups. Finally, communication channels include the methods members of the priority groups normally use to obtain information, for example, bulletin boards and newsletters at senior centers, faith-based organizations, or popular local TV and radio stations.

Step 8: Develop a Plan for Monitoring and Evaluation

This portion of the overall plan is tied directly to the goals that were established in step four: to maintain a kit, abandon total reliance on first responders, understand risks, and believe that preparedness is beneficial to individuals and the community. Measurements of the success of these goals can be defined as output (e.g., activities such as hours of training that are related to the campaign), outcomes, such as measurable changes in the priority groups' goals in step four, and impact measures; in our case, the increase in the numbers of members in the priority who are now prepared. It is important to complete this step (as with the others) in sequence because the costs of monitoring and evaluation must be included in the budget or they may fall by the wayside. Being able to quantify program outputs, outcomes, and impacts is critical for grant reporting and future funding proposals.

Step 9: Establish Budgets and Find Funding Sources

Once all of the components of the plan are in place, along with their associated costs, the total estimated costs can be compared with available funding. At this point, adjustments to the budget can be made and the final draft incorporated into the planning document.

Step 10: Complete an Implementation Plan

Lee and Kotler (2011) describe how this final step insures accountability because it "specifies *who* will do *what*, *when*, and for *how much* [italics in original]" (49). They emphasize that this section is sometimes considered the true marketing plan and suggest that it can even be used by itself to convey the basics of the project with stakeholders (49). Because we intend the social marketing public emergency preparedness process to cover the three recommended actions of making a kit, making a plan, and staying informed, the implementation plan should include at least an outline of those upcoming activities.

Use of Social Marketing in Emergency Preparedness

Although social marketing is mentioned in the emergency preparedness literature (e.g., the PDP model; Coppola and Maloney 2009, 9–13), it is not commonly used to encourage the public to become prepared for emergencies. As would be expected, most emergency managers are not familiar with social marketing principles and so, as Littlefield et al. (2010) note "they did not think that promoting emergency preparedness included listening to key publics, researching barriers to having emergency kits, developing campaigns to address these barriers, and then assessing quantitatively the outcomes of their efforts, as seasoned public relations practitioners would" (256).

Kotler and Lee (2008) described the social marketing process for public emergency preparedness in Canada from 2006–2007 as outlined by Mintz and Wooldridge (2008). I have paraphrased their report extensively because their work is directly applicable to using such a process in the United States. According to the Public Safety Canada website, the process "was created in 2003 to ensure coordination across all federal departments and agencies responsible for national security and the safety of Canadians" (http://www.publicsafety.gc.ca/cnt/bt/index-eng.aspx). Despite major storms and flooding, Public Safety Canada found that the Canadian public remained largely unprepared for emergencies, so they launched a social marketing campaign in 2006 entitled "Is Your Family Prepared?" Research conducted prior to the campaign determined the focus of the initiative and strategies. The program had three purposes: to raise awareness of all hazards, to increase the number of families who have an emergency plan, and to increase the number of people who have a seventy-two-hour emergency kit (156).

The target audience was "parents thirty-five to fifty-three, living in urban centers, skewed towards women" (157). Then in 2007, the program was expanded to include those with lower-income, seniors, and persons with disabilities. The private sector, first responders, and educators were also involved. Their behavioral objectives were to increase the number of people who completed the three purposes previously mentioned. The knowledge objective was to increase the number of people who knew about emergencies and how to respond to them. The belief objectives sought to reduce the number of Canadians who thought that there were some emergencies that you cannot prepare for, but also to increase the number of people who had a sense of urgency about preparation and realized it was necessary to protect themselves and their families (157).

Mintz and Woolridge (2008) noted four main barriers preventing Canadians from taking action: "psychological beliefs that deter vigilance and preparation," not knowing where to obtain information, "reliance on governments in emergency situations" and "language barriers among Canadians who speak neither official language (French and English)" (157). Neither time nor money was found to be significant barriers to preparation; even the lowest income group was willing to purchase a sixty-dollar emergency kit.

In addition to the prepackaged emergency kit just noted, Public Safety Canada also had a website, a toll-free number, and a guide for family preparedness. All but the kit were free. The emergency kits were developed and distributed through a collaboration between not-for-profits and retail chain stores throughout Canada (157). Public Safety Canada used responses to a question about trusted sources in the original research to determine where to distribute the information and materials. Promotion was done through advertising, staff seminars, testimonials, contests, direct marketing, training of government staff, via members of Parliament, and through national marketing and exhibits (158–159).

The budget for the first year was three and a half million dollars. A variety of quantitative measures (visitors to websites, downloads, subscribers to online plans, e-mail subscribers, hotline calls, number of guides ordered) during the first year indicated positive participation. Participation was also tracked during the campaign to determine increases in knowledge and specific emergency preparedness activities (160).

There are a few examples of social marketing for emergency preparedness in the United States. Marshall et al. (2007) described how the Rhode Island Department of Health "used a social marketing approach to identify the public's 'wants and needs' in anticipation of an emergency" (47). They conducted focus groups and in-depth interviews to identify information needs, preferred formats, trusted sources, and other aspects of emergency preparedness. From this research, they developed a detailed thirty-two-page informational brochure prior to launching their "Make a Kit, Make a Plan, Stay Informed" campaign. A postintervention evaluation "indicated that an estimated ten percent of the population changed their behavior by engaging in one or more of the preparedness activities" (47).

In 2012, the state of Hawaii received FEMA funding for its Regional Catastrophic Preparedness Grant Program (RCPGP) to use social marketing to increase public preparedness. As noted on the campaign's fact sheet, the state 1) used scientific market research to determine current public disaster preparedness, 2) used that data to develop a pilot education campaign about emergencies, 3) tested the messages and methods of communication during the pilot campaign, and with that information 4) created a model campaign for the state (Hawai'i Disaster Preparedness 2012, 1).

Day (n.d.) and Walker (2005) studied social marketing for public emergency preparedness as students at the National Fire Academy. Day, a member of the Fire Department and Office of Emergency Management (OEM) in Bellingham, Washington, conducted a full social marketing planning process between 2008 and 2009 with other local agencies and a consulting firm to determine how to encourage seniors to store a seventy-two-hour supply of water. Unfortunately, Day notes that OEM lacked staff to implement the plan.

Walker, a member of the City of Lagrange, Georgia, Fire Department, conducted a literature search on social marketing, completed a community preparedness survey, and did personal interviews as preliminary steps toward a public disaster preparedness campaign. Walker (2005) noted that "interestingly enough, only the public health group was familiar with social marketing or its concepts.... It appears that it will be necessary to educate the emergency management personnel in the concepts of social marketing prior to initiating any kind of preparedness campaign" (35). That is one purpose of this book.

Conclusion

This chapter included a broad overview of theories of behavioral change in preparation for an in-depth explanation of social marketing. The basic theories were: the Extended Parallel Process Model, Stages of Change/Transtheoretical Model, health belief model, theory of planned behavior, and protection motivation theory. Three theories of change specific to natural hazards were also discussed: the Person-relative-to-Event model, the Critical Awareness Model, and the Protective Action Decision Model. The history and process of social marketing was followed by an example using preparation of a seventy-two-hour emergency kit as the social change goal. The chapter ended with several published uses of social marketing for public emergency preparedness in Canada and the United States.

References

Andreasen, A.R. 1995. *Marketing social change: Changing behavior to promote health, social development, and the environment*. San Francisco: Jossey-Bass.

———. 2002. Marketing social marketing in the social change marketplace. *Journal of Public Policy & Marketing*, 21 (1): 3–13.

Azjen, I. 1991. The theory of planned behavior. *Organizational Behavior and Human Decision Processes*, 50: 179–211.

Bandura, A. 1986. *Social foundations of thought and action*. Englewood Cliffs, NJ: Prentice Hall.

Cheng, H., P. Kotler, and N.R. Lee. 2008. *Social marketing for public health: Global trends and success stories*. Sudbury, MA: Jones and Bartlett.

Coppola, D.P. and E.K. Maloney. 2009. *Communicating emergency preparedness: Strategies for creating a disaster resilient public.* Boca Raton, FL: CRC.

Dalton, J.H., M.J. Elias, and A. Wandersman. 2001. *Community Psychology*. Belmont, CA: Wadsworth.

Day, A.D. n.d. *A social marketing approach to disaster preparedness education.* Retrieved on 7/7/13 from http://www.usfa.fema.gov/pdf/efop/efo43356.pdf.

Dooley, D., R. Catalano, S. Mishra, and S. Serxner. 1992. Earthquake preparedness: Predictors in a community survey. *Journal of Applied Social Psychology*, 22: 451–470.

Drabek, T.E. 2013. *The human side of disaster.* 2nd ed. Boca Raton, FL: CRC.

Federal Emergency Management Agency. 2006. Citizen Corps personal behavior change model for disaster preparedness. *Citizen Preparedness Review*, Issue 4. http://home.citizencorps.gov/downloads/pdf/ready/citizen_prep_review_issue_4.p.

Grothman, T. and F. Reusswig. 2006. People at risk of flooding: Why some residents take precautionary action while others do not. *Natural Hazards*, 38 (1–2): 101–120.

Hawai'i Disaster Preparedness. 2012. Hawai'i Disaster Preparedness Pilot Campaign Fact Sheet. http://www.getreadyhawaii.org/index.php/media/hawaii-disaster-preparedness-pilot-campaign-fact-sheet

Kahneman, D. 2011. *Thinking fast and slow.* New York: Farrar, Straus and Giroux.

Kotler, P. and N.R. Lee. 2008. *Social marketing: Influencing behaviors for good.* 3rd. ed. Thousand Oaks, CA: Sage.

Kotler, P. and G. Zaltman. 1971. Social marketing: An approach to planned social change. *Journal of Marketing*, 35: 3–12.

Lee, N.R. and P. Kotler. 2011 *Social marketing: Influencing behaviors for good.* 4th. ed. Thousand Oaks, CA: Sage.

Lefebvre, R.C. 2013. *Social marketing and social change: Strategies and tools for health, well-being, and the environment.* San Francisco, CA: Jossey-Bass.

Lefebvre, R.C. and P. Kotler. 2011. Design thinking, demarketing and behavioral economics: Fostering interdisciplinary growth in social marketing. In *The SAGE Handbook of Social Marketing*. G. Hastings, K. Angus, and C. Bryant, eds., 80–94. Thousand Oaks, CA: Sage.

Ling, J.C., B.A.K. Franklin, J.F. Lindsteadt, and S.A.N. Gearon. 1992. Social marketing: Its place in public health. *Annual Review of Public Health*, 13: 341–362.

Littlefield, R., K. Rowan, S.R. Veil, L. Kisselburgh, K. Beauchamp, K. Vidoloff, M.L. Dick, T. Russell-Loretz, I. Kim, A. Ruvarac, Q. Wang, T.S. Hoang, B. Neff, T. Toles-Patkin, R. Troester, S. Hyder, S. Venette, and T.L. Sellnow. 2010. "We tell people. It's up to them to be prepared." Public relations practices of local emergency managers. In *The Handbook of Crisis Communication*. W.T. Coombs and S.J. Holladay, eds., 245–260. New York: Wiley-Blackwell.

Marshall, R.J., L. Petrone, M.J. Takash, S. Sansonelli, M. Wah-Fitta, A. Bagnall-Degos and A. Novais. 2007. Make a kit, make a plan, stay informed: Using social marketing to change the population's emergency preparedness behavior. *Social Marketing Quarterly*, 13 (4): 47–64.

Mintz, J.H. and T. Woolridge. 2008. Is your family prepared? Public Safety Canada (2006–2007). In *Social marketing: Influencing behaviors for good.* 3rd ed. P. Kotler and N.R. Lee, eds., 156–60. Thousand Oaks, CA: Sage.

Mulilis, J-P. and R.A. Lippa. 1990. Behavioral change in earthquake preparedness due to negative threat appeals: A test of protection motivation theory. *Journal of Applied Social Psychology*, 20 (1): 619–638.

National Academy of Sciences. 2012. *Disaster Resilience: A National Imperative.* Washington, DC: National Academies.

Paek, H-J., K. Hilyard, V. Freimuth, J.K. Barge, and M. Mindlin. 2010. Theory-based approaches to understanding public emergency preparedness: Implications for effective health and risk communication. *Journal of Health Communication*, 15: 428–444.

Paton, D. and J. McClure. 2013. *Preparing for disaster: Building household and community capacity*. Springfield, IL: Charles C. Thomas.

Paton, D., L. Smith, and D. Johnston. 2005. When good intentions turn bad: Promoting natural hazard preparedness. *Australian Journal of Emergency Management*, 20 (1): 25–30.

Prochaska, J.O. and C.C. DiClemente. 1982. Transtheoretical therapy: Toward a more integrative model of change. *Psychotherapy: Theory, Research and Practice*, 20: 161–173.

Rogers, E. 1995. *Diffusion of innovations*. 4th ed. New York: Free.

Thaler, R.H. and C.R. Sunstein. 2008. *Nudge: Improving decisions about health, wealth and happiness*. New York: Penguin.

U.S. Department of Health and Human Services. 2008. *Making health communication programs work: A planner's guide.*

Walker, B.E. 2005. *Increasing citizen disaster preparedness with social marketing*. Retrieved on 7/7/13 at http://www.usfa.fema.gov/pdf/efop/efo38590.pdf.

Witte, K. 1998. Fear as motivator, fear as inhibitor: Using the extended parallel process model to explain fear appeal successes and failures. In *The handbook of communication and emotion: Research, theory, applications, and contexts*. P.A. Andersen and L.K. Guerrero, eds., 423–450. New York: Academic.

3

IDENTIFICATION OF PRIORITY GROUPS

> Seeing people as creative actors in their own lives, and not passive recipients of their fate—or what others "do for them"—sets the tone for how we approach them.
>
> *(Lefebvre 2013, 159)*

This chapter focuses on two important aspects of social marketing research: identification of priority groups and the initial conversation with those participants to determine their learning goals. This field research took place in the same Midwestern county that is the focus of the research in the rest of this book. Discussions with emergency management professionals included their perspectives on priority groups and more general information, such as why people in the local community were not prepared for emergencies, how that preparation could be increased, and identification of community networks to engage more people in the process.

Ultimately, two priority groups were selected for the initial social marketing effort. Three focus groups were held with members of those groups in community settings where they frequently congregate. The results of the interviews with local emergency management professionals are presented first, followed by summaries of the focus group discussions. The chapter concludes with the next steps that will be taken based on that information.

Field Research with Local Emergency Preparedness Experts

Although many emergency preparedness professionals might believe that they lack the resources to undertake any kind of marketing research, Lee and Kotler (2011) cite Andreasen's (2002) *Marketing Research that Won't Break the Bank* with some encouragement to do so. Andreasen suggests using data that is already available, rather than attempting to take on a research project (75). For our purposes, such data might be available from the local emergency management agency or Red Cross chapter, the census bureau, social service agencies (e.g., data about senior centers or people with functional needs), or companies who do commercial marketing. He also noted that systematic observations, even in some cases by volunteers, can provide valuable information (108). He uses the example of systematically visiting beaches at specific times to determine how many children were wearing life vests.

Another method well worth the effort is to pilot test any potential social marketing process (120). An example of that is contained in this chapter—focus groups with seniors and people with functional needs. One final example is quota versus probability sampling (167). This involves sampling specific portions of a population (e.g., seniors living independently, in assisted living, and in nursing homes), and then projecting those results onto the larger population, although this must be done carefully. Lee and Kotler (2011) suggest possibly adding a few questions to surveys done for other purposes and also asking universities for assistance from students or faculty (100).

In order to identify the priority audiences for emergency preparedness training: those most at risk, those most ready for training, and those most crucial to the success of the project, seventeen people who were or had been active in the emergency preparedness field were interviewed. These included people with a variety of backgrounds that often reflected a lifetime of service, including firefighting, hazardous material education and response, law enforcement, emergency medicine, emergency management in higher education, environmental health, epidemiology, service on the Local Emergency Planning Committee (LEPC), education, elected office, Community Emergency Response Team (CERT) coordination, Red Cross staff and volunteers, functional needs services, state emergency management, the military, and municipal disaster services. We knew most of these people from our prior work and identified others by asking participants during the initial interviews. This is commonly known as the snowball technique.

We received a small grant from our university for this project. The interview protocols for the emergency managers and for the focus groups were approved by the university's Institutional Review Board. All of the participants consented to be tape-recorded and the conversations were transcribed. *Each interview with an emergency preparedness professional began with a question about the person's background in the field.* As noted earlier, many people had a diverse background that had evolved through time.

The second question had several parts. We asked, *"As you know, a large percentage of the public is not prepared for emergencies (i.e., does not have a basic seventy-two-hour disaster kit, family communication plan, or information about how to prepare for risks and emergencies in their community). Why do you think some people prepare while others do not? How do you define preparation—these three basics: the seven Public Readiness Index actions (having a supply kit, a portable supply kit, a communications plan, a meeting place, having practiced or drilled, volunteered to help, or taken First Aid within the past five years) or some other way? Are there other skills you would like people to have?"* There were a variety of answers to these questions, so common responses are grouped together in the following section.

Why People Do Not Prepare

The most common response to this question was that people do not prepare because they have not experienced a disaster. Perception of risk was the second factor but it had several aspects. First was that disasters are not common in this region of the country, so the perception that 'it won't happen to me' is in fact somewhat accurate. On the other hand, some people feel that they won't be affected even if a disaster occurs, or worse, that they are adequately prepared when in fact they are not.

One emergency manager stated:

> People have different levels of preparation—they believe they are more prepared than they are. Number one it's an illusion in their mind, kind of like lying to yourself: that one, it can't happen and two, if it does I'll be fine because I'm prepared. In actual conditions, you may not even be able to get out of a simple crisis, like getting out of a house in a fire. . . . Your brain is in primitive mode. . . being able to handle situations that you're not in every day can be problematic under emotional and stressful conditions.

A few professionals noted that some people actually do not know how to prepare or simply do not have the resources to do so. One person noted, "You can't tell somebody to build a kit who can't even fill their refrigerator." Another said, "Low income people are too busy keeping a roof over their heads and dealing with the everyday." An invisible barrier to preparedness is, "There's a lot of mental illness in society"—"people don't wanna think about that kinda stuff when they're battling their own little personal crisis all the time . . . it's overwhelming for them to just get through their days."

Some people believe that the government will take care of them. One first responder stated, "I think people have higher expectations, and I think partly we're responsible for that. We give people a false sense of security in that way too, 'Oh we can handle anything that is thrown at us.'" Two people suggested that "there's the culture of entitlement. I think it's kind of a, 'it's your job to save me' attitude and 'it's not my job to do your job.'" Another person stated, "If you do any studies on police calls, fire calls, you're gonna see that people call for the most, smallest, things, 'I've got a splinter in my finger' or 'I hurt my back.'"

How Do You Define Preparation and Are There Other Skills People Should Have?

The overwhelming response was, as one professional stated, "The four basics (Note: FEMA recently added 'get involved' to the three basics listed in the question.) are sufficient. The key is to get people to do it." Several people emphasized CPR and other basic skills such as, "People should know CPR/AED, how to use a fire extinguisher, basic survival skills, and have a communication network in place including opting in to public notification. People should also have a comprehensive plan for their pets." Another added, "Know how to turn off utilities, do basic first aid and advanced first aid, and be CERT trained." Two professionals suggested that, "People need to be mentally/psychologically prepared and also physically prepared, having kits and an awareness of their environment. They need to have good relationships with others before an emergency."

The third question was "*How can we increase the number of people who prepare (as you have defined it above)?*" One first responder said, "I think the people of this region have had too much sensationalism by the media and they just don't see it happening. But I think people would get prepared if the message and the threat were imminent. We have to refine the message and stay consistent. And get the cooperation of the media to quit sensationalizing every snow storm, every thunderstorm." Two others mentioned cell phone problems. One suggested, "Local emergency managers should modify what is in the kit based on local conditions. Right now the new number one priority in an emergency is 'How do I charge my cellphone?' Now you know to include cell phone chargers and extra batteries and minute cards or whatever."

One long time educator gave this detailed response: "You need to reach out, not to the general public, but to target audiences: age groups, socioeconomic groups, demographics. Customize the presentations; break them into segments, for example, seniors who are immobile and alone. Have an icebreaker. Accommodate for learning style, etc. Use Train the Trainer courses. Train people about diversity, not just how to accept it but how to approach it."

Fostering a Culture of Preparedness

Several people stated the importance of training children, but one emergency manager emphasized the importance of a long-term focus on them:

> Public health taught the kids to cough in their sleeve and they taught the adults. Every kid knows stop, drop and roll. You teach why it's important to have a kit and reinforce that for 18 years and they're going to fear being without one. Adults are not easy to change. Focus the money back downward and let the impact go up. But we're always looking for instant change.

The most thoughtful FEMA director would say, "Ok, my goal is in fifty years to have an America that's more prepared and how do we strategically do that?" Let's be in it for the long haul and not for the short political gain which is what you see most often. Instead "Ok, we're diverting all our preparedness money down to the pre-K and K level and to the grade schools." Obviously you're making a twenty year investment at that point, but you're gonna see the results of that undeniably in my opinion. We're just doing it backwards. I think it requires a rethinking from a political standpoint and the current senior emergency managers who are in positions of power and influence to just say, "Ok we're not doing this right." But you know, the schools will push back and say, "We only have this many hours and we don't have time for emergency preparedness."

But if you start it at the lobby level and a federal legislation level, how they'll do it is like they force you to do everything else. You attach it to money. And then you say, "Well you shall make two hours available in your curriculum for disaster preparedness if you want to receive federal funding for your other programs." But again it has to be a commitment from everybody. The only way I see it ever changing is having a true culture of preparedness. The other cultures that have a typical higher preparedness level are ones that do mandatory military service.

Somewhat echoing the emergency manager's sentiments above, a fire chief described a successful emergency preparedness program they launched in his school district and the need for federal support for such programs:

> One of the first things I did when I got here was to hire a fire education specialist that goes into the schools. She's in there once a week, kindergarten through fifth grade. She teaches not only the fire safety programs that are supported by NFPA [National Fire Protection Association] and the United States Fire Administration, but also a kind of childhood preparedness: what to do in the event of a bad snow storm. She teaches things about Halloween, about unsafe ice. The kids love her and as they go through it, she utilizes a specially trained dog that is very good with kids, and she dresses him up. This is our third year in the program so those kids that we had in the kindergarten grades are now into the second grade, third grade. And the message is sticking to them. I think we have to use the same approach in getting people prepared for emergencies, start teaching at an early age. Not necessarily kindergarten through fifth grade but maybe fourth through eighth and start getting them to understand the potential risk of a tornado, and then have them take that message home to their parents. Using that along with the PTA, and almost conditioning them to understand the risks associated with this and what they can do. That's the only approach I can think of that would work, because you can't teach adults.

> But I think we need to push this onto FEMA to help support it at the federal level. The governor's office, the state EMA and the county, we had the same argument when we came here. But when my fire education specialist sat down with the program and showed them, at least one teacher at each grade level bought in on trial basis. Once she started teaching and they saw that the message was being widely accepted by the students, they allowed her to expand the program and now the kids look forward to her being there. And we have the support of principals. We're part of a three community school system. The two chiefs from the neighboring communities, wanted to be a part of it because we share a fire safety trailer.

Another first responder noted,

> In our area, one of the ways you are providing the messages is through Citizen Corps groups, through CERT and the Medical Reserve Corps and Fire Corps. They're also a good way to get the message to their neighbors 'cause that's the whole essence of CERT. The easy answer would be to get more money because that's always the answer to every problem (laughs). But, you know if we could literally walk door to door and hand people a preparedness kit, that would be one way to do it. But, then how do we get them to maintain that kit?

One emergency preparedness manager emphasized the importance of individual interactions:

> I think it has to be community based. We've seen lots of initiatives that have come out from the CDC, from FEMA, the billboards saying "Are you prepared? Get prepared." But unless you know someone who has gone through an emergency and maybe tells you about it, it doesn't really hit home. So I think hearing from your neighbor or your friends or your relatives, that's probably the best approach, that personal interaction. And having them hear it from first responders, they really add a lot of credibility. And we can't forget the public officials, although their role is different. It's not about our government being prepared. It's really individuals who make up a community. It just all has to be coordinated.

One public official noted, "Well I think we need to probably do even a better job of letting people know. After the last blackout when we were out of electricity, people said 'Why aren't you notifying us?' We responded, 'Well if you had Nixle you would have gotten the message but if you don't give us your email address or your phone number, we can't notify you.'"

Several preparedness educators gave detailed responses. One said, "Know your audience, how to persuade them, and how they prefer to learn. Social media and short bites are becoming the norm. Also, provide individual classes to groups and communities; short, humorous PSAs for TV, Twitter, FB and other social sites. Produce regular seasonal articles on preparation; contests with preparedness prizes; preparedness contests between companies; celebrities putting their clout behind projects. For seniors, incentives help and teaching children."

Another educator responded, "Provide clear and consistent messages and visuals; civic info; social media, repeated often; and corporate sponsorship (especially from a trusted source). Give information to pet owners. Go to the schools. Implement the American Red Cross pillowcase project. Go to the churches. And make it inexpensive, Walgreens—one item per month, and field test it." A third educator emphasized, "Demonstrate how easy it is. You Tube videos can be positive. Get the biggest bang for the buck. Don't guilt trip people."

A fourth emergency manager suggested, "Go for the low hanging fruit. The American Red Cross has several apps to help people be prepared. Go to the grassroots: faith based organizations and block clubs. Get everyone together who does preparedness, build on synergies and identify the gaps."

The fourth question, about identifying priority groups, had multiple parts. It was worded as follows: *"We are trying to segment the general public into smaller groups in order to improve our training effectiveness. One way this is typically done in public health education is to identify 1) those who are at highest risk, 2) those who are most ready for the training, and 3) those who are most crucial for the success of the training. For example, in our community, we might identify seniors in high rises as most in need of education on evacuation and shelter-in-place. Could you please suggest a population for each of the three groups?"*

People Who Are at Highest Risk

The two groups that many of the respondents identified as being at greatest risk were seniors/elderly and people with functional needs. The third group was children, followed by the poor. Other groups mentioned more than once were the mentally ill, people with pets, and the socially isolated. One manager's response was common, "The ones at highest risk are those dependent on others: children, the elderly, people with access and functional needs, people with pets, people with low socioeconomic status, lower education and those with no support system."

Another emergency manager said:

> I think the elderly because our communities continue to age. And people that don't have transportation. You say "OK this shelter's open and you can come here and stay warm and get some food." Well they don't have a way to get there. And it's those that are economically disadvantaged and those that are disabled. Those are the people who are most at risk. The addiction and mental health agency actually works with some of the homeless populations too. There's some cross links, they work with the shelters. It's a very complex network.

One emergency manager made a direct link between lack of preparedness and mental illness:

> I think people are pretty unanimous on who's at highest risk so I don't think I'll provide anything intuitive here other than saying make sure mental health is on the top of that list. I would love to see somebody go back and do a study of the people who were in the Superdome during Katrina and see how many of them self-report mental health problems like depression. My sense is that you're probably going to find a pretty high correlation between people who had a past or current history of mental health problems and why they ended up there and didn't evacuate or weren't better planners.

A service provider during emergencies added:

> Those with functional needs, that's been more of our focus in terms of preparedness. Looking at populations that are marginalized: the elderly, children for sure, people with mental health issues, visually impaired, hearing impaired, people that can't get places 'cause they don't have a car, where transportation's an issue, foreign language, all those things. At any given time any one of us may have one of those functional needs. If we blow out our knee and we're on crutches, we're gonna have difficulty moving around. Making sure that people are thinking, "I'm OK today but you know, our neighbor across the street, she's not healthy at this point" and maybe you look after them.

Those Who Are Most Ready for Training

The most common response to this question was children, followed closely by people who had recently been through an emergency and seniors. One person pointed out that those who have heard the message before are receptive because they realize that they need to take action. Many people also suggested specific organizations or groups, such as Scouts, businesses, school personnel, elected officials, government employees, block watch groups, nonprofits, churches, poor women, doctors, Offices of Aging, health departments, and pharmacists.

Two responses focusing on children were notable. The first was:

> The kids are the most receptive because it's easy to make it fun and enjoyable to them, not so much to an adult. Kids carry that message for the rest of their lives. It's a safety language. We do all the traditional stuff still, "stop, drop and roll" and "call 9–1–1" but there's no reason you can't add "build an emergency kit" in there too. So you have them make it a fun kit of toys first and say, "What's your favorite toy you'd like to take? Is your blanket

> in here?" Then you build up to "Now we have to take it more seriously and here's your real items you would need in an emergency." You start out with the fun things they would want to take if you had to leave the house, without scaring them.

The second longer response related to children was, "The most ready are children at community centers during vacations and summer camp and those who take the Red Cross's Masters of Disaster preparedness module. The can also be taught in after school programs and at church outside of school hours. The families should be involved."

As one person noted, people who have heard the message before may be ready for training:

> I think it's one of those things where you know you have to take action. Maybe they have been told by their social worker or visiting nurse, "Hey, let's make sure you have some extra medication on hand." And that triggers in their mind, "Well if I need medication I probably need to have some other things." I ask, "So what do you think can happen if you don't have electricity?" When they hear it like that they realize I'm not talking about a 9/11 type of emergency. It's simple stuff that will happen every year. It's carbon monoxide.

Those Most Crucial for the Success of the Training

The most frequent response to this question was that the Police Chief, Fire Chief, and elected or public official are the most crucial for success because they have the most influence on the community. Furthermore, as one manager noted, "It is in their best interest." The second most common responses were: those who already do emergency preparedness such as CERT and state and local emergency management agencies, schools, support agencies for people with functional needs, and youth. One person also said, "I think your community leadership and their buy-in. And community isn't necessarily like a city, but the faith-based community or social groups. If you have a leader in that group who says, 'This is important folks, we all need to do this,' I think that goes a long way." People also mentioned corporate and foundation donors, people who manage high rises and can facilitate having preparedness training onsite, heads of households, and church leadership.

Regarding those who already are involved in preparedness, one person gave this response:

> This is a great opportunity for CERT to get more involved. That whole dynamic just gets people thinking. The more people hear it from someone that they trust, versus the government, the better. If people talked about the

> actions that they're taking more, that would do it. But this is so difficult now because there's just so many things going on in this world. I think people have lost focus on what's really important. You know, you don't need those 80 things on that list. You pick out the ones like, "OK I could definitely use that. If I was in a bind those things would be helpful."

The fifth question was, "*Do you know how we can reach members of these groups?*"

The majority of responses to this question focused on those most at risk. Regarding senior citizens, one person said, "Focus on reaching people through their own networks like senior centers and also word of mouth through friends versus the government and also through CERT members." Others added social service agencies and through doctors. Several people grouped seniors and people with functional needs together. An example of this was this suggestion, "The hardest are the ones who aren't mobile. Go to the recreation center and families of residents. Have families encourage people to attend: senior caregivers (family, county workers, facilities), kids and parents, seniors (independent or with caregivers). Also those who self-identify as having functional needs to their health and services department, EMS or fire."

Children were also considered a group at high risk. One fire chief noted, "We spent two years with a local church doing their emergency action plan for their facility. And we're now going back because they want us to train their staff in the event that they had children in the school and there was an emergency." Others recommended going to the elementary schools and high school health classes, youth summer school, the YMCA, Scouts, and using the Red Cross pillowcase project.

Question six was, "*Do you have any stories of when public preparation was successful?*"

Unfortunately, there was only one local example of successful preparedness, but it was a memorable one recounted by a fire chief:

> I'll go back to when I was chief in another city. We had been teaching a fire education program in the schools for a number of years. And this was around 2007–2008 because they used it for the basis of a grant. We had a single mother who needed to pick some extra hours up so she took her daughter with her and they were driving to the office. It was the day after Christmas, the 26th. Her boss had given her some extra work to bring home and she told the boys to clean up the house. There were three boys, if I remember correctly, and so the boys did. They packed up all the wrapping paper. They packed up everything else in the plastic bags. The mother was a smoker and they emptied the ash trays into the plastic bags too and they threw them in the garage. And in about a relatively short period of time the house started filling up with smoke. But the oldest boy, who was eleven years of age, had been through our safety program, did the right thing. He

> took the younger boys outside the back door, across the street to a meeting place that they had determined, and then got a neighbor to call 9–1–1 from outside. So we had three kids exit the home safely. The fire department was called. The fire was contained to the garage. Obviously there was smoke damage into the house but all three kids got out of the house safely. It literally mimicked exactly what we talk about in exit drills in the homes. So there's a good example of how things improve.

Our final question was "*Can you suggest other people that we should interview?*" We were able to interview some of the people that were suggested and noted the rest for future research.

Focus Group Results

As noted at the outset of the chapter, three focus groups were held with two priority groups identified through the interviews with local emergency professionals. Although children were identified as a group at high risk, due to time and resource constraints, we chose to focus on seniors and people with functional needs. The focus group protocol and questions were approved by the university's Institutional Review Board. Funds from the small grant we received were used to purchase gift cards from a local grocery and snacks for the participants. They all consented to be tape-recorded and the conversations were transcribed.

Two focus groups were held at senior centers located on opposite sides of the city. The third focus group was with people with functional needs at a nonprofit agency that helps people to live and work independently. One senior center site was identified by one of our colleagues in local government. The other was selected due to its location and designation as a center where seniors could obtain free meals, making it a busy, popular site. Both center managers were enthusiastic about hosting a focus group and actively recruited participants. The nonprofit agency serving people with functional needs also was very enthusiastic about hosting a focus group due to the advocacy of one of our colleagues who introduced us to the executive director and a senior staff member.

The questions for the focus groups were much different from those for the interviews with the emergency managers. We were attempting to begin to identify what people in these groups knew about preparing for emergencies, what they wanted to know, and how they would prefer to learn. The responses from both types of groups are summarized after each question. The first question was an icebreaker. It was "*First, we'll have people introduce themselves (first name only) and share why they are interested in talking about emergency preparedness.*" A member of one senior group responded, "Because it's the theme of the Commission on Aging forum in a couple of weeks."

Another person answered passionately:

> I came here because I'm concerned about this community. Not just this community, but we are living in another world today. I think people in this country are so laid back, they have no clue what to do, when to do it or how to do it. I grew up in a military family. I lived in occupation in Japan for two years. I've seen devastation. The apartments here that seniors are living in, if something happened there they don't even know where the exits are, they have no clue. So I do work with CERT and I received my thirty-five year award as a Red Cross volunteer two years ago. I'm just concerned about what I see. I know what to do but all the people on my street, they have no clue. If anything happens, if the lights go out, they get devastated.

A third person stated, "I was more in tune when I had small children. Now that everybody's grown I let it go by the wayside. You have been the protector all this time. Well, now all of a sudden, you're out of a job."

The functional needs group, as would be expected, gave somewhat different responses. One advocate stated, "I'm interested because one of the committees I'm on is working on making the public aware of people with functional needs and what to do when there's an emergency. We did a presentation to a lot of people that work in emergency preparedness. Two of us also dealt with emergency shelter preparation in 2012 or 13."

FIGURE 3.1 Training together reinforces new behaviors. (Image courtesy of 4774344 sean/iStock)

Another person said:

> I've always been interested because me being a person with a disability, I wanted to know how things worked. We work on the second floor and if something goes wrong, if there's an emergency, how are they gonna help us get down to the first floor? 'Cause I know the fire department would bring us down but with disabilities, throwing me over their shoulder, that's not going to work. Our chairs are upstairs. What will we sit in when we get downstairs?

One participant gave a much longer response:

> I just want to say first of all I wanted to know about emergency preparedness because I'm curious about the first responders, the fire fighters, the EMTs, do they have any disability awareness or sensitivity training that they have to go through at all? And then I wanted to share a situation. I will be living by myself because my two daughters are going to be moving out pretty soon. Well in the winter time I actually walked out to my mailbox with my white cane and because there was snow on the ground I got totally disoriented so I was spinning around in the middle of the street and finally someone came running out to me, one of my neighbors. So when you talk about a buddy system and knowing your neighbors, that's the best thing you could ever do.

Other people talked about their personal living situations and their concerns about getting help, not only in emergencies, but also in daily life. One young man said, "You know, the more I think about all this, don't be surprised if you see me start carrying a little bag on the back of my chair with a bunch of information and a little pouch on me in case something happens.

The second question was "*What type of emergencies do we have in our county?*" People in both groups identified emergencies common to our area but focused primarily on power outages and severe weather. The third question was, *"Have you prepared for emergencies in some way? What have you done?"* One senior woman responded, "I have water, I have candles, I have a lantern and I fill my bathtub if I have to. I keep flashlights and batteries. And then we have a CERT backpack that has minimal stuff in it but I put more personal things in it too. Get an LED lamp. You should make copies of your driver's license, bank cards and those types of things and have it in an extra wallet. You also need your medication." Another offered, "Well, something I've gotten in the habit of doing lately is every three days I put gas in my car, so the tank is always full." But another person candidly said, "Nothing. I'm gonna be honest, nothing." One man who shared stated, "I am more alert. I listen for information on TV and on the radio. I have kitty litter in my car trunk to use for traction. I got a kit from the Senior Day we had recently.

My medication and water. Cooler with ice packs for insulin. A flashlight that you can shake to light it. Extra flashlight batteries."

The fourth question was, "*What would be the best way for you to learn; for example, would you like to read some information, make preparations in a group, purchase a seventy-two-hour supply kit of food, etc.?*" Participants in all three groups said they preferred to learn together, except for gathering private information, such as a list of their medications, and making preparations in their homes, such as storing water and food. One senior added, "At the AARP meeting the person from the health department told everybody there about putting that medications list and all your contact numbers in a Ziploc bag that goes on the outside of the refrigerator. Also the Vial of Life. I think if you leave it up to the individual they're more apt to set it aside and forget about it."

We then distributed FEMA brochures about preparedness for seniors to that group. The nonprofit serving people with functional needs had received a donation of Go Stay Kits® that we gave to the participants along with FEMA brochures. We gave them a few minutes to look through the information. No one had any questions about the information.

The final question was "*If you were going to take one small step to becoming prepared, which of the activities would you do first and why?*" The most common responses were that people would either replace the water they had set aside or add water to their 'to go' kit, primarily to take medication. Next people said that they would add either a first aid kit or other items to their existing kit. Some frequent answers were a flashlight and spare batteries, snacks, chocolate or candy, and an emergency crank radio that could also charge a cell phone. Some other items included "putting things in a tote bag with wheels that had my name and address on it," a change of clothes and shoes, a blanket, and cat food.

One woman in the functional needs group surprised us all by saying, "I would put in alcohol. I was thinking whiskey. Well actually you could get hurt really bad and need to be numbed from the pain." We reminded her that she could not take alcohol into a shelter, but that prompted someone else to say, "I have a question. How do we get to the shelters?" Because people lived in different places, we suggested that they should find out how they could get to a shelter and make that part of their personal emergency plan. One person had a longer list, "I would add water, snack crackers and extra medication, bleach and trash bags, insect repellent, a whistle, a blanket and change it with the season, toilet paper and towels, butterfly bandages with a first aid kit, my Swiss army knife, peroxide, wines [probably added due to the woman saying she would add whiskey], a waterproof container with matches and emergency phone numbers."

Two people in the functional needs group had already taken some special precautions. One man said, "I put all my important papers that I need, birth certificate, social security card all that stuff, insurance, the whole thing, in one pouch and I have the pouch where I can grab it and take all my important papers with me." Another woman said that both her and her boyfriend, "have all of our things in fireproof boxes. I actually keep mine, which sounds kinda strange, in my main bathroom of my home because to me, we're more likely to have a tornado and that's my safe room."

We concluded the focus group by distributing the gift cards but also offering the participants further training the following year. Nearly everyone in all three groups indicated they were interested by entering their contact information on a sign-up sheet. We were able to offer this opportunity because in the course of doing the formative research described in this chapter, we obtained another small grant to continue working with the priority groups.

Conclusion

One of the most crucial steps in the social marketing process is the identification of priority groups. This chapter summarized interviews with local emergency managers to obtain their input on three types of training participants: those most at risk, those most ready for training, and those most crucial for the success of the intervention. Although children ranked high in the first and second categories along with senior citizens, we chose to focus on seniors and those with functional needs because it became apparent that educating children was a given and should be a standard part of the school curriculum as fire prevention is today. During two focus groups with senior citizens and one with people with functional needs, we began to learn what participants knew about emergency preparedness, whether they had taken any action, and how they wanted to learn more. The results of those conversations will be used to inform the content and process used in actual training and follow up. Now that the priority groups have been identified, in the next chapter we will prepare an in-depth social marketing plan for public emergency preparedness.

References

Andreasen, A.R. 2002. *Marketing research that won't break the bank*. San Francisco, CA: Jossey-Bass.

Lee, N.R. and P. Kotler. 2011. *Social marketing: Influencing behaviors for good*. 4th ed. Thousand Oaks, CA: Sage.

Lefebvre, R.C. 2013. *Social marketing and social change: Strategies and tools for health, well-being, and the environment*. San Francisco, CA: Jossey-Bass.

4

DEVELOPING A DETAILED SOCIAL MARKETING PLAN

> Social marketing practice is guided by ethical principles. It seeks to integrate research, best practice, theory, audience and partnership insight, to inform the delivery of competition sensitive and segmented social change programmes that are effective, efficient, equitable and sustainable.
>
> *(A Consensus Definition of Social Marketing endorsed by the Boards of the International Social Marketing Association, European Social Marketing Association and the Australian Association of Social Marketing, 5 October 2013)*

In this chapter, with the permission of the authors and Sage, we will use Lee and Kotler's (2011) Social Marketing Planning Worksheets (as published in *Social Marketing: Influencing Behaviors for Good*) to formulate a generic plan for applying social marketing methods for public emergency preparedness training. They can be found in that book as Appendix A (459–471) and more in-depth information about each step can be found within its chapters. For our purposes, I have included information about public emergency preparedness from my local community that can be easily modified to fit your particular situation or focus.

Lee and Kotler note that their ten-step plan is similar to a typical marketing effort but that it differs in three ways. First, because in social marketing our efforts are focused on a priority group, the group is identified *before* we identify specific behaviors that we intend to foster. Second, the competition is not specified in the situation analysis because we have not determined the behavior we want to encourage. Third, in social marketing, goals are the quantifiable outcomes of the process (e.g., the increase in the number of seniors and people with functional needs who have an emergency communication plan) rather than the overall purpose of the plan (42–43). Lee and Kotler's Worksheet information is distinguished from the author's words in the following sections either by being in bold print or by number, for example, 1.1.

Step 1: Provide Background, Purpose, and Focus for Plan

Organization/Partners: county Office of Emergency Management, local Red Cross Chapter, state Emergency Management Agency

1.1 Summarize key background information leading to the development of this plan.

The plan was developed due to a lack of public emergency preparedness repeatedly demonstrated through national surveys and behavior before, during, and after emergencies; the effects of climate change, including an increase in the number and intensity of severe storms and temperatures; droughts and wildfires; demographic changes that have put more people at risk for all hazards (e.g., an increase in the number of senior citizens); and geographic movement of populations to locations that are more vulnerable to natural emergencies, among other variables.

1.2 What is the campaign *purpose*, the intended impact (benefit) of a successful effort?

The purpose of the campaign is increased public preparedness for all hazards. Despite an increase in natural and humanly caused disasters and the events of September 11, 2001, in the United States, only approximately one-third of the public is prepared for emergencies by having taken three basic actions. These consist of being informed (learn what protective measures to take before, during, and after an emergency), making a plan (prepare, plan, and stay informed for emergencies), and building a kit (build a kit for disasters to be prepared) (Ready.gov 2014).

1.3 What is the campaign *focus*?

The campaign focus is on taking the three basic public preparedness actions.

Step 2: Conduct a Situation Analysis (Identify 2–3 Bullet Points for Each)

Internal Factors: Organizational Resources, Expertise, Management Support, Internal Publics, Current Partnerships, Distribution Channels

2.1 What internal *strengths* will your plan maximize?

- Community partners and supporters
- Functional disability experts and consumers
- Ongoing county Office of Emergency Management trainings

2.2 What internal *weaknesses* will your plan minimize?

- Little prior experience with social marketing processes
- Lack of paid staff to implement social marketing process
- Limited financial resources

External Forces: Cultural, Technological, Demographic, Natural, Economic, and Political/Legal Forces and External Publics

2.3 What external *opportunities* will your plan take advantage of?

- Emergency management community member awareness of the social marketing process due to interviews during formative research
- Increased awareness of more severe weather events due to climate change
- Ability to refer to new county preparedness website within social marketing process to inform and educate participants

2.4 What external *threats* will your plan prepare for?

- Lack of a history of large-scale local disasters fosters public complacency
- Competing priorities: focus on first responder resources versus training citizens to be self-sufficient

Prior and Similar Efforts

2.5 What findings from *prior and similar efforts* are noteworthy, those of yours or others?

The use of social marketing for public emergency preparedness has rarely been documented in the literature. In at least two instances, the process could not be completed due to lack of staff or funding. It will be important to plan for the long term to insure that resources are available for the duration of this initial project, including evaluation and follow-up. This is also emphasized in the social marketing literature.

Step 3: Select Target Audiences

3.1 Describe the *primary target audience* for your effort. Consider variables including size, demographics, values and lifestyle, geographics, related behaviors, and/or readiness to act.

There are two priority audiences. The first is senior citizens (those sixty-five years of age and older), who in 2012 comprised approximately 16 percent of the population (200,841 people) of our county. More specifically, we are focusing on seniors who are able to frequent senior centers and have a local community group there where we can meet with them and follow up with the emergency preparedness training. Formative research has indicated that seniors in this group are eager to learn about preparedness and wish to do so in a group setting. We can identify the senior centers through local municipalities, our regional area agency on aging, and the county Division of Senior and Adult Services. Seniors are particularly at risk during emergencies because many have functional needs (e.g., diminished mobility, sight, hearing, cognition; reliance on medications for daily living; and the use of oxygen or other medical supports that require electricity). They also may be socially isolated or become so quickly due to a medical condition or loss of spouse or other family members.

The second priority group is people with functional needs. We do not know the actual number of people with functional needs because they need to self-report this information. However, one large agency serving adults and children with developmental disabilities currently has 13,000 clients and another that helps people with functional needs to live and work independently has 1,700 clients. Functional needs is a broad definition that includes people who "may have physical, sensory, mental health, and cognitive and/or intellectual disabilities affecting their ability to function independently without assistance" (FEMA 2010). We will identify members of this community through social service agencies and the city and county public health departments who do outreach to this community. People with functional needs are at higher risk not only during emergencies, but also in daily life because of their need for assistance.

3.2 If you have *additional important target audiences* that have influence on your target audience, describe them here to keep them in mind as you develop strategies. They may end up being messengers or distribution channels.

Additional priority audiences for both of these groups are the staff and volunteers at the senior centers, the support agencies, and the AARP (formerly the American Association of Retired Persons), as well as similar groups of people with functional needs and their advocates. Family members of the seniors and people with functional needs are also important for their role in helping their relatives to maintain the kit, communication plan, and resources that they receive during the social marketing effort.

Step 4: Set Objectives and Goals

Objectives

4.1 *Behavior Objective*: What, very specifically, do you want to influence your target audience to *do* as a result of this campaign or project?

We want the priority audiences to complete the three emergency preparedness actions recommended by FEMA (Ready.gov 2014). The first is to be informed so that people become aware of the hazards that may impact the area where they live and other places they may visit. Once they recognize these hazards, they need to know how to physically take care of themselves before, during, and after each type of emergency, for example, by evacuating or sheltering-in-place.

The second action is to make a plan so that people know before an emergency what they will do to remain safe, how they will communicate with family and friends, and then reunite with them afterward. Such plans should be made for the home, workplace, school, and other locations where people may be during an emergency.

The third action is to build a kit that includes basic necessities for seventy-two hours during and after an emergency. We want the participants to begin by gathering items for a basic kit including water (one gallon per person per day); food (non-perishable) and a can opener if necessary; a first aid kit and prescription medicines; a battery or crank operated radio; a flashlight with extra batteries; a whistle to call for help; a dust mask and plastic sheeting and duct tape for sheltering-in-place; moist towelettes and trash bags with ties for sanitation; a wrench or pliers to turn off utilities; and local maps. We will encourage participants to add other items recommended by FEMA and specific to the season or type of hazard they are most likely to face.

4.2 *Knowledge Objective*: Is there anything you need them to *know*, in order to act?

Knowledge in this case includes information or facts that the priority audiences must have in order to carry out the behavior. The priority audiences need to know the hazards common to their region and other possible risks. They need to know how to obtain information in an emergency, how to take action to protect themselves, the elements of a communications plan, and how to prepare a seventy-two-hour kit.

4.3 *Belief Objective*: Is there anything you need them to *believe*, in order to act?

Beliefs are attitudes or feelings that the priority audience may have about the behavior. One common belief people have about emergencies and disasters is that "It won't happen to me." Another is that first responders will be immediately available to help in an emergency. We need the priority audience to understand that emergency preparedness can prevent them from injury or death. It can also help them to be relatively comfortable during many types of emergencies, including loss of electricity and when they are unable to leave their homes, such as during snowstorms. Personal preparation can enable people to help others directly and also allows first responders to assist those most in need.

Goals

4.4 What quantifiable, measurable goals are you targeting? Ideally, these are stated in terms of *behavior change*.

Lee and Kotler (2011) emphasize, "you should recognize that what you determine here will guide your subsequent decisions regarding marketing mix strategies.

It will also have significant implications for your budgets and will provide clear direction for evaluation measures later in the planning process" (44).

The first short-term goal is to increase awareness of local hazards and how people with functional needs and seniors might prepare for them. Public service announcements will be used on the most popular radio and TV stations. Notices with a similar message will be distributed to agencies who serve people with functional needs, to senior centers, and to the additional priority audiences noted above.

The next step is the actual training. The initial outreach will be to a total of 300 people: 200 seniors and 100 people with functional needs. A two-hour training will be conducted with 10 groups of 20 people at the largest senior centers around the county. The behavioral change goal is very modest. If all the trained seniors complete the three preparedness actions, one thousandth of the 200,000 seniors would be more prepared. Another goal is to have 50 of the 200 trained seniors, or 25 percent, serve as ambassadors to other senior centers or members of their communities in subsequent trainings and outreach efforts. A second follow-up meeting will take place with the original participants two months later to determine the actions that they have taken and what additional support they might need to complete them.

A two-hour training will also be conducted with ten groups of ten people with functional needs at agencies or facilities where this community routinely comes together for services or resides. The behavioral change goal for this group is also very modest. We hope to have twenty-five people serve as ambassadors to

FIGURE 4.1 Learning in priority groups promotes social change (Image courtesy of monkeybusinessimages/iStock)

other people with functional needs. A second follow-up meeting will take place with the original participants two months later to determine the actions that they have taken and what additional support they might need to complete them.

Step 5: Understand Audience Barriers, Benefits, and the Competition

Barriers

5.1 Make a list of *barriers* your audience has to adopting the desired behavior. These may be related to something physical, psychological, economical, skills, knowledge, awareness, or attitudes. (Try for a list of 5–10.)

Some of the barriers to taking the three preparedness actions are:

- Lack of a radio or television to obtain information about preparedness actions and training
- Functional needs (e.g., difficulty reading or writing) that make it difficult to prepare a communication plan
- Do not know where to purchase items for the seventy-two-hour kit
- Are unable to pay for items for the kit
- Have an aversion to thinking about emergency situations because of past experiences
- Lack of family or friends to ask for support during an emergency
- Lack of urgency regarding planning due to responsibilities as a caregiver
- Feeling overwhelmed by the planning exercise and lists of items for the kit

Benefits

5.2 What are the key *benefits* for performing the behavior that your target audience wants in exchange for performing the behavior? This answers the question "What's in it for me?" (Try for a list of 2–3.)

Some of the benefits include:

- Protecting oneself from possible injury or death during an emergency
- Peace of mind knowing that one will have adequate medications and supplies during an emergency
- Knowing where to obtain accurate and timely information during an emergency to inform oneself and others

Competition

5.3 What are the major competing *alternative behaviors*?

Some of the competing alternative behaviors for both priority groups are things that they find pleasant or engaging such as games, videos, and interacting with others.

5.4 What *benefits* do your audiences associate with this behavior?

The benefits are companionship and fun, particularly if being with other people is an important part of their day.

5.5 What *costs* do your audiences associate with this behavior?

The majority of the competing activities are free, so the primary cost is giving up some common leisure time activities.

Step 6: Craft a Desired Positioning Statement

6.1 Write a statement similar to the following, filling in the blanks. "We want [target audience] to see [desired behavior] as [descriptive phrase]."

"We want seniors and people with functional needs to see personal emergency preparedness as an important part of keeping themselves and their neighbors safe during emergencies of all types."

Step 7: Develop Marketing Strategies

7.1 Product: Goods or Services

7.1.1 What is the *core* product, the major perceived benefit(s), your target audience wants from performing the behavior that you will highlight? (Choose one or a few from those identified in 5.2.)

The core product is being able to be safe and relatively comfortable during an emergency without having to rely on first responders for help with daily living.

7.1.2 What is the *actual* product, the features/name for the desired behavior? Refer back to your behavior objective in 4.1 and refine/finalize here.

The actual product includes three components: knowing about local hazards and where to obtain information about them before, during, and after an emergency; preparing an emergency plan for all hazards and a communication plan; and preparing and maintaining a seventy-two-hour kit.

7.1.3 Are there any *tangible objects* that will be included in program and campaign efforts?

We will provide handouts about local hazards and how to prepare for them and maintain an e-mail and mailing list to notify participants about upcoming trainings and activities.

7.1.4 Are there any *services* that will be included in program and campaign efforts?

There will be annual updates to the initial training.

7.2 Price: Fees and Monetary Incentives and Disincentives

There will not be any charge for the training. We hope to provide a small financial incentive, such as a gift card, for participating in the full training and follow-up, if grant funding is obtained.

7.2.1 If you will be including tangible objects and services in your campaign, what, if anything, will the target (priority) audience have to *pay* for them?

If the participants do not already have the items for the seventy-two-hour kit, they will need to purchase them. Those items include some or all of the following: one gallon of water per person per day; non-perishable food for three days; a can opener for food if needed; a battery-powered or crank radio and extra batteries; a flashlight and extra batteries; a first aid kit; a whistle to signal for help; a dust mask and plastic sheeting and duct tape to shelter-in-place; moist towelettes, garbage bags, and plastic ties for personal sanitation; a wrench or pliers to turn off utilities; local maps; prescription medications and glasses; food and water for pets; important family documents and copies of insurance policies, identification, and bank account information in a waterproof, portable container; cash in small increments; emergency references such as information from FEMA or a first aid book; a sleeping bag or warm blanket for each person; a change of clothing for the current season; household chlorine bleach and a medicine dropper to disinfect water with instructions; a fire extinguisher; matches in a waterproof container; personal hygiene items; paper cups, plates, and plastic utensils; paper towels; paper and pencil; books or activities.

7.2.2 Will there be any *monetary incentives* for target (priority) markets?

No, unless grant funding is obtained or partner agencies offer them.

7.2.3 Will there be any *monetary disincentives* you will highlight?

Yes, the cost of not preparing oneself and home for emergencies, particularly not having adequate insurance for local hazards such as flooding.

7.2.4 Will you use any *nonmonetary incentives*?

Yes, we would like to publicly recognize the people with functional needs and seniors who have competed the training and also provide them with a certificate or badge.

7.2.5 Will you use any nonmonetary disincentives?

No, negative messages are inappropriate for this type of campaign.

7.3 Place: Making Access Convenient

As you determine each of the following, look for ways to make locations closer and more appealing, to extend hours, and to be there at the point of decision making.

7.3.1 *Where* will you encourage and support your target audience to *perform the desired behavior* and *when*?

We will encourage the priority audiences to perform some of the three behaviors during the initial training. Those behaviors are learning about local hazards and how to prepare for them and developing a family communication plan. We will identify items that should be in a seventy-two-hour kit and encourage people to gather them at home, and then report back during the follow-up meeting two months later.

7.3.2 *Where* and *when* will the target market acquire any related tangible objects?

During the training, the priority audiences will be given handouts about preparing for local hazards, forms to complete a communication plan, and a checklist for preparing the seventy-two-hour kit. They will gather the tangible objects for their seventy-two-hour kit at home. We will suggest where they can purchase items at a reasonable cost, such as a battery-powered radio, reusable water containers, and a basic first aid kit.

7.3.3 *Where* and *when* will the target market acquire any associated services?

The priority audiences will be told how to contact the local emergency management agency or Red Cross chapter if they require assistance.

7.3.4 Are there any groups or individuals in the distribution channel that you will target to support efforts?

Yes, we will identify specific individuals at these agencies who can assist people with their planning.

7.4 Promotion: Persuasive Communications: What Will You Say, Who Will Say It, How, And Where?

Messages

7.4.1 What key messages do you want your campaign to communicate to target audiences?

There are four key messages. First, people with functional needs and seniors are at more risk during emergencies of all types. Second, it is easy to prepare and they can get help. Third, that preparing will bring them and their families peace of mind. And fourth, that if they are prepared they can help their family, friends, and neighbors.

Messengers

7.4.2 Who will deliver the messages and/or be the perceived sponsor?

The county Office of Emergency Management, local municipalities, senior centers, and any other groups or agencies that assist seniors or people with functional needs who are willing to distribute information or co-sponsor the activities.

Creative Strategy

7.4.3 Summarize, describe, or highlight elements such as logo, taglines, copy, visuals, colors, script, actors, scenes, and sounds in broadcast media.

"Seniors, are you ready for the next emergency?" Seniors helping seniors; county READY website logo on all materials (look back to messages above), and the same message for people with functional needs.

Communication Channels

7.4.4 What communication channels will you use?

Radio and TV PSAs; social media (especially for families of people with functional needs and seniors); AARP and secondary audiences (e.g., functional needs service agencies, senior service agencies, families of people with functional needs and seniors, faith-based organizations); county READY website; messages on utility bills; senior volunteer agencies (Senior Corps/Corporation for National and Community Service) and caregiver networks (e.g., Visiting Nurse Association).

Step 8: Develop a Plan for Evaluation and Monitoring

8.1 What is the *purpose* of this evaluation? Why are you doing it?

The evaluation is being done to determine how much behavioral change the training achieved and to improve further social marketing efforts.

8.2 *Who* is the evaluation being conducted for? Who will you present it to?

The evaluation is being completed for the county Office of Emergency Management, the local Red Cross chapter, funding agencies, senior centers, and agencies serving people with functional needs and sponsoring partners. We also will report the results back to the participants and trainers.

8.3 *What goals* from Step 4 will be measured?

We will measure how many of the three preparedness actions the participants completed initially, if they took further steps (such as adding more items to their kit or volunteering to help with emergencies), if they have maintained their kit, and if they have assisted others in preparing.

8.4 *What* techniques and methodologies will be used to conduct these measures?

At the beginning of the two-hour training, we will distribute a pretest to evaluate the level of knowledge the participants have about local hazards and preparing for emergencies, a family communication plan, and the suggested contents of a seventy-two-hour kit. As part of our agreement with the participants, we will have a one-hour, follow-up meeting two months after the initial training. Because self-reporting is generally not an accurate measure, we will ask the participants to bring as much of their information (i.e., communication plan and documents) and their kit (with the exception of water and food) to the meeting as possible. We will have checklists available so that we can count the items the participants have gathered/completed.

8.5 *When* will these measurements be taken?

The measurements will be taken prior to the initial training and at the follow-up meeting. We will contact people who cannot attend the follow-up meeting by phone.

8.6 *How* will measurements be reported and to whom?

The measurements will be reported using a simple comparison of the pre and post training surveys, along with a narrative about the social marketing effort.

Step 9: Determine Budgets and Find Funding Sources

9.1 What costs will be associated with *product*-related strategies?

Lee and Kotler (2011) define product-related costs as those "most associated with producing or purchasing any accompanying *tangible goods* [italics in original] and developing or enhancing associated *services* [italics in original] needed to support behavior change. Costs may include direct costs for providing these goods or services, or they may be indirect costs, such as staff time" (418–419).

The primary tangible goods that will be provided to the participants are training-related materials and the training itself. The costs for printing, etc., may exist in the agency's budget for public outreach. Staff time for training might currently be allocated in the existing budget. This time could also be grant funded or donated by qualified volunteers in the local community. At least initially, the contents of the seventy-two-hour kit will be provided by the participants.

9.2 What costs will be associated with *price*-related strategies?

Recognition programs and rewards will be limited to those that can be given at minimal cost, such as certificates, and to items that may be donated from other agencies or local vendors.

9.3 What costs will be associated with *place*-related strategies?

At least at the outset, the delivery channels will be those places where people with functional needs and seniors already congregate. Mileage costs for trainers will be one minimal cost associated with this phase of the campaign.

9.4 What costs will be associated with *promotion*-related strategies?

For this initial phase of the training, we will use existing communication channels between government and nonprofit agencies and networks.

9.5 What costs will be associated with *evaluation*-related strategies?

As outlined in Step 8 above, staff will be doing the evaluations as part of the training and in the follow-up sessions.

9.6 If costs exceed currently available funds, what potential additional funding sources can be explored?

Given current budget constraints, the most likely source of additional funding is through grants from state or federal emergency management agencies and those serving people with functional needs or senior citizens.

Step 10: Complete an Implementation Plan

10.1 Sample Implementation Plan

Table 4.1 is an example of an implementation plan based on what has been developed in this chapter.

10.2 If conducting a pilot or plan in several phases, complete a grid for each phase.

TABLE 4.1 The social marketing implementation plan

WHAT	*WHO*	*WHEN*	*HOW MUCH*
Increase awareness of local hazards and how seniors and people with functional needs might prepare for them	Staff and community partners through PSAs and info to senior centers and agencies and residences of people with functional needs	First two months of campaign	Cost of staff and community partner time to develop and distribute messages
Outreach to senior centers and agencies/residences of people with functional needs to schedule two-hour training and one-hour follow-up two months later with 10 groups of 20 seniors each	Staff and community partners	Third month of campaign	Cost of staff time
Training and follow up	Staff and community emergency volunteers	Fourth through sixth month	Cost of staff time and materials for training
Evaluation	Trainers and staff	Seventh month to compile and distribute results	Cost of staff time
Ongoing support for emergency preparedness questions	Staff	Ongoing	Cost of staff time

Conclusion

Because social marketing is a carefully designed process of identifying priority audiences and understanding their perspectives and needs, formulating goals and objectives, and developing market strategies, budgets, and evaluation instruments, this chapter provided a detailed example of each step. An implementation plan could be prepared for an entire social marketing intervention or for one phase, such as focusing on one priority group in a specific geographical area for a limited time. With this foundation in the social marketing planning process, we can now explore how we will work with the people in our community. That is the subject of the next chapter.

References

Federal Emergency Management Agency. 2010. Guidance on Planning for Integration of Functional Needs Support Services in General Population Shelters. http://www.fema.gov/office-disability-integration-coordination.

Lee, N.R. and P. Kotler. 2011. *Social marketing: influencing behaviors for good.* 4th ed. Thousand Oaks, CA: Sage.

Ready.gov. 2014. http://www.Ready.gov.

5

WORKING WITH YOUR LOCAL COMMUNITY

> Community preparedness education programs need to be grounded in community development, and focus strongly on supporting existing networks, and encouraging trust between community members, partners and government agencies.
>
> *(Australian Red Cross 2013, 24)*

The goal of community emergency preparedness is resilience during and after emergencies. Individuals with strong ties to many other people will have access to resources that socially isolated people will not. But rather than attempting outreach and education to these individuals, emergency preparedness professionals can achieve exponentially greater impact by focusing on the larger groups to which they belong. Having an ongoing relationship with these groups or social networks is also a vital resource for communication and outreach during all phases of emergency management.

Because emergency management professionals are typically not trained to do community engagement, the purpose of this chapter is to explain some methods of identifying and working with social networks in tandem with segmenting the community into smaller priority groups for focused emergency preparedness training. The chapter begins with a brief history of the concepts of social networks and social capital. That is followed by a discussion of how social capital may be identified and measured. The chapter ends with a local example of how this might be done as part of an emergency preparedness social marketing initiative.

Social Networks and Social Capital

Social networks contain like-minded individuals who trust and communicate with each other frequently and rapidly. These networks contain the social capital that makes them so valuable to individuals. Putnam (2000), the author of *Bowling Alone: The Collapse and Revival of American Community*, defined social capital as "connections among individuals—social networks and the norms of reciprocity and trustworthiness that arise from them" (19). The term social capital was coined in the early 1900s by a school supervisor to emphasize how people helping one another could benefit the whole community. Since then, Putnam and many others have enriched our knowledge about this process. The terms social capital and social networks are sometimes used interchangeably. Aldrich (2012) noted that "researchers have struggled to clarify whether social capital comprises the data about, reputations of, and information flowing between members of a group or if it *is* [italics in original] the network of relationships and connections itself" (29). Regardless of which term is used, the main point is that understanding the relationships between members of a community is vital to fostering social change and in this case, emergency preparedness.

Scholars have identified three types of social capital: bonding, bridging, and linking. Bonding relationships are those between, for example, people in church groups and fraternal or ethnic organizations. Members of these groups rely on each other for reciprocal help and psychological support based on their common characteristics. They are typically exclusive of others. Alternatively, bridging social capital is inclusive, and may reach beyond ethnic, socioeconomic, and other types of networks. Putnam (2000) included "the civil rights movement, many youth service groups and ecumenical religious organizations" (22) in this category.

Linking social capital is a relatively new concept. Aldrich (2012) used Szreter and Woolcock's (2004) definition, which is "networks of trusting relationships between people who are interacting across explicit, formal or institutionalized power or authority gradients in society" (655). He identifies these as relationships that cross vertical gradients, such as with a government official, NGO, or some authority figure (34). As Putnam (2000) notes, "bonding social capital constitutes a kind of sociological superglue, whereas bridging social capital provides a sociological WD-40" (23). Following that analogy, linking is a lifeline, particularly in emergencies when people need resources from beyond their local community.

Having worked for decades to improve public health using social marketing, Lefebvre (2013) emphasizes the importance of not focusing on individual behavioral change. Regarding public emergency preparedness, this means we would seek to work with bridging and linking networks connected to priority groups in order to have the greatest potential impact. Although we focused on those at highest risk in the previous chapter, we would now want to identify and work with networks of people who were already practicing emergency preparedness (e.g., CERT teams) to broaden their impact through their additional community connections, that is, through their bridging and linking community connections.

Obviously, for good or bad, people learn behaviors from those they closely associate with. Just as membership in a social network can be beneficial, it can also be a detriment. Anyone who has attempted to quit smoking, for example, knows that if one's family members and friends continue to smoke, it is much more difficult to stop. This is also true for other unhealthy behaviors. For example, Christakis and Fowler (2007) found that people are more likely to become obese if those they associate closely with become obese. Lefebvre (2013) emphasized that social determinants, such as poverty and lack of access to information, should be recognized and taken into consideration when working with social networks. He also cites Goyal's (2007) research on three factors that influence whether people adopt specific behaviors: the number of connections people have with others, how dense those connections are, and how far removed they are from others. These concepts may sound familiar because they are similar to the "six degrees of separation" theory. With this brief introduction to social capital and social networks, we can begin to apply this framework to preparedness.

Social Capital and Emergency Preparedness

It has been widely recognized that social capital is an important component of successful disaster recovery (Dynes 2002; Norris et al. 2008; Cutter et al. 2008). We'll examine one case in particular because it describes several methods of measuring social capital, all related to post-disaster recovery. Political scientist Daniel Aldrich relocated to New Orleans with his family a few months before Hurricane Katrina struck. They were able to evacuate ahead of the storm, but the experience and its aftermath took his research agenda in a new direction. Aldrich (2012) studied four major disasters: the 1923 Tokyo earthquake, the 1995 Kobe earthquake, the 2004 Indian Ocean tsunami, and Hurricane Katrina in 2005. He discovered that "*high levels of social capital—more than such commonly referenced factors as socioeconomic conditions, population density, amount of damage or aid—serve as the core engine of recovery*" [italics in original] (15). He found that communities that lacked social capital were much less able to recover and rebuild post-disaster. This holds true even from one neighborhood to another, as Aldrich (2012) and Chamlee-Wright (2010) found in researching recovery in New Orleans.

Of course, there are many benefits to having high social capital. It is hard to imagine how a community or nation could thrive without it. This was one of Putnam's concerns. Aldrich states that "First, social ties can serve as informal insurance, providing victims with information, financial help, and physical assistance" (44). "Second, organized communities can effectively mobilize and overcome barriers to collective action (Olson 1965)" (48). "Finally, social networks raise the cost of exit from a community and increase the probability that residents will use 'voice' (Hirschman 1970)" (49). Recognizing that these are important community characteristics, we will now explore how to identify them.

Measurement of Social Capital

When Putnam (2000) began to study the decline in civic participation in associations and other social groups in the last decades of the twentieth century, he was faced with the challenge of measuring events that had already happened. He compares his situation with that of scientists trying to prove the impact of human activity on global warming. He took a similar triangulation approach. Where scientists studied tree rings and pollen in polar ice (26), Putnam used a wide variety of measures and statistically compared and contrasted the records to insure that he was not misinterpreting the information.

Because Putnam was studying change in American social life, he had to use large databases. Those included the National Election Studies (NES), a national survey sample of political behavior conducted nearly every two years since 1952, and the General Social Survey (GSS), which measures social behavior and has been conducted approximately every other year since 1974 (419). Putnam also used the Roper Social and Political Trends data. Between September 1973 and October 1994, Roper did in-person interviews of a national cross section of nearly 2,000 people, 10 times per year, for a total of more than 410,000 respondents (420). He also drew on the DDB Needham Life Style surveys, begun in 1975, which measure social, economic, and personal beliefs and behaviors. In 1999, this database contained over 87,000 responses, and it is still being conducted (Putnam 2000, 420). Finally, Putnam incorporated data from the Americans' Use of Time survey, which contains time diaries kept by samples of Americans in 1964, 1975, 1985, and 1995 (424). While each database had a different focus and some drawbacks, by combining them and using rigorous statistical analysis, Putnam was able to provide a rich picture of the decline of civic participation in the late twentieth century.

Ultimately, he framed this disturbing picture by emphasizing that American community life had once been rich and could be so again, proposing ways for that to occur. At the outset of the book, he tells a true story about two men, John Lambert and Andy Boschma, who lived in Ypsilanti, Michigan. Lambert, a sixty-four-year-old retiree had been on a kidney transplant waiting list for three years when Boschma, then thirty-three years old, found out about Lambert's situation and offered to donate one of his kidneys to him. When they were in the hospital, Boschma told Lambert how much he respected him and that he wouldn't hesitate to do it again. The beauty and power of this story is that Boschma is white and Lambert African-American. What created this powerful bond? They bowled together. As Putnam stated, "In small ways like this—and in larger ways, too—we Americans need to reconnect with one another. That is the simple argument of this book" (28). I have provided this description of Putnam's work to emphasize the importance of social capital for emergency preparedness but also to contrast it with the methods used by Aldrich (2012), who attempted to measure social capital in Japan, India, and New Orleans specifically as it relates to disaster recovery.

FIGURE 5.1 Social capital enhances community resilience (Image courtesy of moodboard/ Thinkstock)

Aldrich's (2012) work was local due to his focus on post-disaster recovery. As noted earlier, he chose four megacatastrophes, categorized as such because over 1,000 people died and thousands of homes and buildings were destroyed (3). Out of the hundreds of disasters worldwide over the past decades, Aldrich chose the 1923 Tokyo earthquake, the 1995 Kobe earthquake, the 2004 Indian Ocean tsunami, and Hurricane Katrina in 2005 based on seven factors.

First, the disasters were from different eras to avoid conclusions that recovery was due to "modern technologies and correspondingly higher citizen expectations" (18). The disasters were also from different nations to rule out culture as the reason for recovery. Third, the cases cover different phases of recovery. In Tokyo and Kobe, the phases were medium and long term, whereas in New Orleans and India, the focus was on the short and medium phases of recovery. The strength and competence of the various governments also varied, as did the types of disasters themselves. The sixth and seventh variations were the types of ethnic homogeneity and the speed at which different neighborhoods recovered within each location (20–21). These variables enabled Aldrich to perform extensive statistical analyses that further distinguished the importance of social capital in post-disaster recovery. Readers who would like to review his methodology may do so in his book and other publications. I offer a somewhat detailed explanation of Aldrich's variables in the cases that follow to heighten awareness of methods to identify social capital in a variety of cultures and situations.

Aldrich identified four methods of obtaining data about social capital in different countries and time periods: attitude surveys like the GSS; behavioral surveys, such as the NES; surveys of types of participation (e.g., voter turnout) and field experiments (40). The 1923 Tokyo, Japan, earthquake destroyed nearly half the capital city, including 350,000 homes. Over 60,000 people died and 1.6 million residents were affected (56). Aldrich used data from thirty-nine police precincts from between 1922–1933 to measure the influence of damage, population density, human capital, economic capital, and social capital on population change. This was possible because local police observers kept extensive records on residents.

Although social capital was the best explanation for the recovery of neighborhoods, it was in Tokyo that the negative side of social capital became apparent. Shortly after the earthquake, due to false rumors that Koreans were committing crimes, such as setting fires, between 2,000 and 5,000 of these residents were murdered by fellow citizens, often with the complicity of the police. In this case, not only did people lacking bonding social capital not have access to resources, but they also lost their lives.

The 1995 Kobe, Japan, earthquake was one of the largest economic disasters in Japanese history. It killed 6,500 of the port city's 4 million people and left over 300,000 people homeless (74). Aldrich analyzed data from Kobe's nine wards consisting of "economic status, dependence on welfare, damage, socioeconomic inequality, and geographic conditions" and found again that social capital was a strong determinant of recovery. In this case, he used the number of new nonprofit organizations (NPOs) created after the earthquake as the measure of social capital between neighborhoods, which often had dramatically different recovery rates (75).

The 2004 Indian Ocean tsunami killed more than 230,000 people in Indonesia, India, Sri Lanka, and Thailand (91). Aldrich focused his research on the fishing villages in the Tamil Nadu region of India. He used three types of data: qualitative case studies of six affected villages in the region, quantitative data of about 62 villages, and data about 1,600 individuals (92). The case studies consisted of interviews with 27 people (using a translator) in five of the six villages regarding their experiences. These were follow-up interviews with a few of the people who participated in the 1,600-person research. Information about the sixth village came from secondary and tertiary sources. The villages were chosen to represent a range of social capital and recovery. Aldrich also was able to use eighty anonymous survivor interviews taken by the Indian government after the disaster. The data from this phase of the research provided a rich picture of how people relied on others during the recovery (95).

Within the villages, a major source of both bonding and linking capital were the urr panchayats or village councils. They played a key role in identifying the dead, conveying community needs to aid workers, and then delivering that aid within the villages. Communities without urr panchayats were not able to obtain aid due to their lack of organization. However, the urr panchayats also prevented

"outsiders" such as widows, Dalits, and Muslims from sharing in the post-tsunami assistance. This is another instance of the dark side of social capital that Aldrich identified in analyzing the Tokyo earthquake. Because the Indian government could not possibly interview each individual, it gathered data from religious groups but this meant that people not affiliated with a denomination were left without resources.

Another source of data about social capital was based on Louis's (2005) quantitative information about three types of aid people received from the Indian government via NGOs in sixty-two marginalized villages. That aid included days spent in a relief camp, the number of eligible families who received a package of initial supplies, and the number of families who received 4,000 rupees (Aldrich 2012, 106). Louis's variables included measures of wealth, connections to the outside world, location and caste (109). Aldrich's statistical analyses of the data revealed exclusion from aid based on caste, while the wealthy were more likely to receive assistance, as were people who owned their homes (117).

The third quantitative database that Aldrich analyzed was prepared by Sawada, Sarath, and Shoji (2006). It is called the University of Tokyo—Tamil Nadu Agricultural University (UT-TNAU) Household Survey on the Tsunami-Affected Villages (Aldrich 2012, 120). The research team for this study interviewed 1,600 people and gathered information about "demographics, socioeconomic status, tsunami damage, caste, social capital, and recovery outcome" (Aldrich 2012, 120). Linking social capital was measured by the amount of aid people obtained from the government, from NGOs, or both. Bonding social capital was determined by a measure typical in India and Southeast Asia: attendance at weddings and funerals, specifically "the amount of money given at weddings" and the number of funerals attended prior to the tsunami (122).

Statistical analysis of the data yielded some anticipated outcomes; for example, older, more educated and married people received more aid, as did those who lost family members (123). But an unexpected result was that both bonding and linking social capital "highly determined levels of aid" (Aldrich 2012, 123). This is because these activities strengthen existing relationships and forge new ones. Based on his work in Tamil Nadu, Aldrich makes several suggestions for policymakers, some of which will be echoed in the discussion of New Orleans that follows. First, evacuation planners should strive to keep existing neighborhoods intact so that people can help each other to recover and hopefully return home together (127). This is particularly important for those who are alone or socially isolated. His second recommendation was to find ways to include the groups that may have been excluded due to cultural or social factors by increasing their bonding and linking capital (128). Finally, Aldrich suggests that those who are distributing aid have "comprehensive profiles of affected areas" so that governments and relief agencies do not allow people in need who lack social capital to be left without necessary resources (2012, 128).

Aldrich's (2012) fourth and final case was post-Katrina recovery in New Orleans. In 2005, Hurricane Katrina killed 1,500 people, displaced 2 million, destroyed 300,000 homes, and caused $130 billion in damage (133). Recovery in some areas of the city has been complete, whereas other areas languish to this day. Aldrich proposes that this is due to the types of social capital that were present before the storm. He contrasts two neighborhoods to highlight this situation. The Village L'Est community in northeast New Orleans, although poor, had both bonding and linking capital that enabled it to achieve "a 90 percent population recovery and business recovery within two years after the disaster (LaRose 2006; Faciane 2007)" (Aldrich 2012, 130). Alternatively, Aldrich proposed that the Lower Ninth Ward failed to recover because it lacked linking capital (2012, 130). In this case, Aldrich used turnout of eligible voters in the 2004 Presidential election as a proxy for social capital. He notes that this method had been used before in the United States (see Verba and Nie 1972; Hamilton 1993; Walsh and Warland 1983; Putnam 2000; Aldrich and Crook 2008) (61).

After the disaster, although it was well understood that temporary housing was critical, many people adamantly opposed the siting of FEMA trailers and modular homes in their neighborhoods. Aldrich found that "those localities with more politically active and involved citizens who voted in past elections . . . received the fewest trailers" (134). In order to rule out other factors that could also influence the trailer siting decisions, Aldrich reviewed the scholarship on five variables (technocratic criteria, discrimination against minorities, socioeconomic criteria, amount of damage, and civil society) that are typically used to identify locations for contentious facilities or what he calls "public bads" (136). He included all the potential zip codes within the potential siting area and incorporated measures for each variable.

To measure technocratic criteria, he calculated the area of the zip code block and its population density. Discrimination against minorities was measured by the percentage of nonwhite residents within the zip code. Aldrich measured socioeconomic criteria using "income, unemployment, percentage of the population with income below the poverty line, education level and house prices" (137). The amount of damage was measured by floodwater depth in each zip code. Lastly, civil society was measured by voter turnout, as previously mentioned, by laying the precinct data over the zip code blocks (140). Please see Aldrich's (2012) book for a detailed discussion of the data analysis.

The results showed that areas with higher voter turnout received far fewer trailers than those with low turnout (143). With all of the other variables set at their means, areas where close to 80 percent of people voted would receive fewer than 100 trailers. However, areas with less than 30 percent turnout, would receive up to 1,200 (144). Aldrich emphasized that both businesses and governments routinely consider social capital when deciding where to locate problematic facilities (see Aldrich 2008; Hamilton 1993; Sherman 2006).

Although Aldrich's (2012) extensive research focused on the importance of social capital for post-disaster recovery, I have included it here because the social capital that a community has during the preparedness phase will be what they have when disaster strikes. In daily life, these bonds are what make a neighborhood or a city one's home. Aldrich's advice to emergency planners is as applicable to preparedness as it is to recovery. He calls for "a reorientation of disaster recovery programs at all levels—local, national, and international—away from the standard fixes focused primarily on physical infrastructure and toward social infrastructure" (151). He says that our current focus in disaster planning and recovery is similar to decades of Western thinking regarding foreign aid. Now entities such as the World Bank recognize that building trust and other types of social infrastructure is as important as financial investment (158). This means that we should seek to maintain relationships between people just as we attempt to provide them with food and shelter, including during evacuations (157).

At the local level, Aldrich suggests building relationships with schools, faith-based organizations, and other places where people traditionally build social networks, and emphasized the importance of assisting in their development where they are lacking. He identified a few organizations that have recognized the importance of social networks for emergency preparedness. One is the United States Department of Agriculture, which has included social capital in helping communities prepare for wildfires (164). Another program that is directly applicable to the focus of this book is Seattle Neighborhoods Actively Prepare (SNAP). This is a step-by-step plan for preparing neighborhoods by having people get to know one another, build plans together, exercise them, and take action as a community when disasters occur, all the while maintaining linking social capital to their local Office of Emergency Management (Aldrich 2012, 164). The program is similar to the way CERTs are formed and operate. As Aldrich (2012) reiterates, "ensuring that social capital is on the agenda will create future plans that generate efficient and effective recovery and build resilient communities" (166).

A Strength-Based Approach

Typically we plan for social change by identifying the problems with individuals, groups, or organizations, rather than looking for what is right with the person or system. A strength-based approach is very different from the usual public policy and change initiatives, particularly in the United States. For example, we rarely ask survivors of a natural disaster how they managed to do that. Why were they prepared? Did friends or neighbors help them? Did they have prior experience or training? Instead, we create a list of deficiencies and system failures with specific fixes for each one. While problems cannot be ignored, focusing on what we know works allows us to build on solid ground rather than speculation. This process is known as appreciative inquiry (AI) (Cooperrider and Whitney 2005).

AI is a continuous five-phase cycle (i.e., the 5-D cycle) that engages a whole system, such as a group, organization, or company, to identify a positive way to implement a change (Cooperrider, Whitney, and Stavros 2003; Barrett and Fry 2005). It has been used worldwide in the public, private, and nonprofit sectors. Sometimes the need for change is driven by a crisis, but generally it is done to improve service or to prompt innovation. In the first phase (Define), organizational stakeholders identify an "affirmative topic" or goal. In our case, for example, we would invite a few people in the functional needs priority group who have been preparing for emergencies to be core stakeholder partners. Together we would prepare our question, for example, "how can we insure that people with functional needs have the resources and support they need during emergencies?" To answer the question, we would talk with additional people who have been participating in emergency preparedness to find out directly from them what they have done and how to reach others who are not yet prepared. This second step is known as the Discovery phase.

During the third phase, we would compile all the results and identify common themes (Dream). Some themes might be the inherent resilience that people with functional needs develop due to overcoming challenges in daily life over time. Others might be the importance of a network of family and friends, and of a way to communicate with them under emergency conditions. From those themes, stakeholders would formulate an implementation plan (Design) that members of the organization affirm and commit to because it includes their input. This is a crucial component of the process, because it insures the buy-in of the members of the organization and therefore dramatically improves the potential success of the intervention. The final implementation phase is known as Destiny, because the plans are put into action. The cycle then is repeated and refined as the organization moves toward completing the change.

Although appreciative inquiry is a highly intentional and focused process, I have explained it briefly to suggest its compatibility with the social marketing process and to advocate for its strength-based, stakeholder approach. Lefebvre (2011) noted that using an asset-based approach aligns with two basic values of social marketing because it focuses on the needs of the priority group and is a co-creative process rather than a program imposed from outside the community (39).

In social marketing, focusing on people who are not yet in an at-risk situation is called an upstream approach. In contrast, a downstream approach is one in which we are pulling people out of the stream. Because we are planning for public preparedness, I'll use an example about flooding. Using an upstream approach, we would take some of the steps advocated by the National Academies. These might include educating the public so that they would reject living in an area prone to flooding and insuring that people had adequate insurance for their property and its contents in case flooding did occur. A downstream approach addresses those who have already been affected, such as those who live in flood prone areas and have experienced losses.

During the field research discussed earlier in the book, local emergency preparedness professionals identified those at greatest risk as seniors and people with functional needs. Children were also considered at high risk, and we noted that educating children should be a given, just as fire prevention awareness is routinely taught in the schools now. Building on our earlier discussion of social capital and now also using a strength-based approach, how can we identify local social capital specifically for people with functional needs and seniors to help these groups prepare for emergencies?

Measures of Social Capital Based on Specific Circumstances and Goals

Social capital has been used in a wide variety of disciplines to measure both positive and negative aspects of social activity. Van Deth (2008) conducted an exhaustive review of the research on social capital in an attempt to identify a common definition for the term that could be operationalized in order to find standard variables that were scientifically valid. He compared research that measured structural and cultural aspects of social capital along with individual and collective measures using both quantitative and qualitative methods (160). First he found that most researchers were concerned about the many definitions of the term to the point that some thought it was used too commonly to be valid. However, that did not prevent scholars from using various combinations of variables in an attempt to quantify it. The problem with developing a precise definition, Van Deth emphasized, was that what social capital is depends on the circumstances under study and the function it is thought to perform. We saw this earlier in this chapter with Aldrich's (2012) application of a variety of measures of social capital related to post-disaster recovery. Van Deth ultimately concluded that, because the term social capital is ambiguous and refers to specific circumstances, a "bottom up" approach to identifying measures is acceptable, because there is no alternative (154).

In terms of public emergency preparedness, to reach the largest number of people with each effort it makes sense to focus on the intersection of four types of groups: the most at-risk priority groups; those with high bridging and/or linking social capital, those who are already involved in emergency preparedness, and those who vote. These four groups were identified based on what the research suggests and our unique need to prepare the public as efficiently and effectively as possible. First, those most at risk are the priority groups for training. Next, as discussed earlier in this chapter, there are three types of social capital: bonding, bridging, and linking. The reason to not reach out to bonded groups, for example, churches and fraternal and ethnic organizations, is because they are typically small and by their nature are exclusive. Third, people who are already practicing emergency preparedness provide a positive example for others and are a rich resource for community outreach and education. Finally, people who vote are typically more socially engaged than those who do not, as has been demonstrated repeatedly in prior research. The following table provides examples of each type of group.

Using the method outlined in Table 5.1, we would determine where the highest concentration of members of each priority group could be found (e.g., senior centers and associations of people with functional needs). Because those groups connect individuals to outside resources, they facilitate bridging social capital. However, it is best to have both bridging and linking social capital. Organizations that provide the latter are listed in the third column (e.g., Area Agencies on Aging and governmental service providers). Groups practicing emergency preparedness (e.g., CERTs and local emergency management agencies) are identified in the fourth column. As these groups work with the senior and functional needs organizations, they are broadening their own bridging and linking social capital, thus increasing the impact of their service. The geographical area of each of the three groups can then be matched with the locations of the bridging social capital sites to facilitate interaction and training opportunities.

Finally, given that voter turnout information is a widely recognized measure of social capital and is easily available from local Boards of Elections, areas of high turnout can be laid over the other information to identify geographical areas with this additional strength. As mentioned earlier in the chapter, Aldrich (2012) laid precinct data over zip code blocks to make identification even easier. The Australian Red Cross (2013) advises that, "through the recognition of existing social capital resources, communities could use key agents within at-risk populations and work within existing programs to provide integrated and engaging messaging on preparedness"

TABLE 5.1 An example of local social capital identification

Priority Group	*Bridging social capital*	*Linking social capital*	*Groups practicing emergency preparedness*	*Voting turnout*
Seniors	Senior centers Retirement communities Assisted living	Area Agencies on Aging City and county service agencies AARP	CERT volunteers American Red Cross (ARC) volunteers Local/state emergency management agencies	Board of Elections
People with functional needs	Associations of people with functional needs Apartments designated for people with functional needs Group homes	City and county service agencies Nonprofits who serve them	Same as above	Same as above

(17). Yet in our example, we are not just preparing messages, but also building on existing social capital through interactive training and relationship building.

Conclusion

Emergency management professionals typically have minimal resources for public preparedness education, so outreach efforts need to be focused effectively to achieve the greatest benefit. The concepts of social capital and social networks were introduced in this chapter as a means to contact larger groups of people who already have existing, strong relationships with one another. Members of these groups have trusting relationships with each other and credibility that government agencies sometimes lack, particularly in times of crises. The types of social capital: bonding, bridging, and linking, were then discussed because it is the latter two that are particularly important in recovery from disaster. That section was followed by a variety of social capital measurements used specifically for studying why some communities rebuild and thrive post-disaster while others fail to recover.

An additional resource for community engagement was provided through a description of a strength-based approach to organizational and social change, appreciative inquiry. That process is similar to social marketing in its positive focus and co-creative methodology. The chapter closed with an example of local social capital identification based on two priority groups: seniors and people with functional needs. Potential bridging and linking networks were proposed along with groups practicing emergency preparedness as resources for outreach and education.

Emergency management professionals may be very interested in working with their communities but lack the time, personnel, or resources to pursue some of the methods described thus far. The next chapter provides some alternative ways to staff and fund public preparedness projects so that communities and regions can become more resilient.

References

Aldrich, D.P. 2008. *Site fights: Divisive facilities and civil society in Japan and the West*. Ithaca, NY: Cornell University.

———. 2012. *Building resilience: Social capital in post-disaster recovery*. Chicago, IL: The University of Chicago Press.

Aldrich, D.P. and K. Crook. 2008. Strong civil society as a double-edged sword: Siting trailers in post-Katrina New Orleans. *Political Research Quarterly* 61 (3): 378–389.

Australian Red Cross. 2013. *Relationships matter: The application of social capital to disaster resilience*. Carlton, Vic, Australia: Australian Red Cross.

Barrett, F.J. and R.E. Fry. 2005. *Appreciative inquiry: A positive approach to building cooperative capacity*. Chagrin Falls, OH: Taos Institute.

Chamlee-Wright, E. 2010. *The cultural and political economy of recovery*. New York: Routledge.

Christakis, N.A. and J.H. Fowler. 2007. The spread of obesity in a large social network over 32 years. *New England Journal of Medicine*, 357: 370–379.

Cooperrider, D., D. Whitney, and J.M. Stavros. 2003. *Appreciative inquiry handbook: The first in a series of AI workbooks for leaders of change*. Bedford Heights, OH: Lakeshore.

Cooperrider, D. and D. Whitney. 2005. *Appreciative inquiry: A positive revolution in change*. San Francisco, CA: Berrett-Koehler.

Cutter, S.L., L. Barnes, M. Berry, C. Burton, E. Evans, E. Tate, and J. Webb. 2008. A place-based model for understanding community resilience to natural disasters. *Global Environmental Change—Human and Policy Dimensions*, 18 (4): 598–606.

Dynes, R. 2002. *The importance of social capital in disaster response*. Newark, DE: Disaster Research Center.

Faciane, V. 2007. Vietnamese community thriving in eastern New Orleans. *New Orleans Times-Picayune*, 23 April.

Goyal, S. 2007. *Connections: An introduction to the economics of networks*. Princeton, NJ: Princeton University.

Hamilton, J. 1993. Politics and social costs: Estimating the impact of collective action on hazardous waste facilities. *RAND Journal of Economics*, 24 (1): 101–125.

Hirschman, A. 1970. *Exit, voice, and loyalty: Responses to decline in firms, organizations, and states*. Cambridge, MA: Harvard University.

LaRose, G. 2006. Asian businesses drive eastern New Orleans recovery. *New Orleans City Business*, 2 October.

Lefebvre, R.C. 2011. Social models for social marketing: Social diffusion, social networks, social capital, social determinants and social franchising. In *The SAGE Handbook of Social Marketing*. G. Hastings, K. Angus, and C. Bryant, eds., 32–43. Thousand Oaks, CA: Sage.

———. 2013. *Social marketing and social change: Strategies and tools for health, well-being, and the environment*. San Francisco, CA: Jossey-Bass.

Louis, M. 2005. *Study on discrimination and exclusion in state relief*. Madurai, India: People's Watch-Tamil Nadu.

Norris, F., S. Stevens, B. Pfefferbaum, K. Wyche, and R. Pfefferbaum. 2008. Community resilience as a metaphor, theory, set of capacities, and strategy for disaster readiness. *American Journal of Community Psychology*, 41 (1): 127–150.

Olson, M. 1965. *The logic of collective action: Public goods and the theory of groups*. Cambridge, MA: Harvard University.

Putnam, R. 2000. *Bowling alone: The collapse and revival of American community*. New York: Simon and Schuster.

Sawada, Y., S. Sarath, and M. Shoji. 2006. Household Survey on the Tsunami Affected Villages. University of Tokyo—Tamil Nadu Agricultural University (UT-TNAU Dataset).

Sherman, D. 2006. *Not here, not there, not anywhere: The federal, state and local politics of low-level radioactive waste disposal in the United States*. Paper presented at the 2006 Northeastern Political Science Association conference, Boston.

Szreter, S. and M. Woolcock. 2004. Health by association? Social capital, social theory, and the political economy of public health. *International Journal of Epidemiology*, 33 (4): 650–667.

Van Deth, J.W. 2008. Measuring social capital. In *The handbook of social capital*. D. Castiglione, J.W. Van Deth, and G. Wolleb, eds., 150–176. New York: Oxford University.

Verba, S. and N.H. Nie. 1972. *Participation in America*. New York: Harper and Row.

Walsh, E.J. and R.H. Warland. 1983. Social movement involvement in the wake of a nuclear accident: Activists and free riders in the TMI area. *American Sociological Review*, 48 (6): 764–780.

6 ALTERNATIVES TO TRADITIONAL RESOURCES

> The journey of a thousand miles begins with a single step.
>
> *Lao Tsu*

Attempting to use social marketing as the foundation for improving public emergency preparedness requires resources that go beyond what is available to most emergency management professionals, whether in local governments or nonprofits. This chapter responds to that need by describing a variety of other mechanisms for obtaining personnel, expertise, and funding to build viable and lasting social marketing programs. The chapter begins with some guidelines for using social media to work with priority audiences. That is followed by a bit of encouragement and advice for beginning a social marketing program. It then proceeds with information on finding volunteers and maintaining a volunteer program in local emergency management agencies. The important topics of alternative funding sources and determining a return on investment for a social marketing program complete the resources offered in this portion of the book.

Effective Use of Social Media

Lefebvre (2013) has studied social media tools, including "blogs, podcasts, wikis, social network sites, apps and text and multimedia messaging" and insists that they should be an important component of any social marketing program, providing they are used as a means to interact with people, rather than to simply broadcast messages (415). They are important because so many people are now engaged in communities and conversations online; and if we want to facilitate behavioral change, we need to be able to enter discussions pertinent to that behavior.

Lefebvre (2013) advises us to become "collaborators, conveners, facilitators, brokers and network weavers (Lefebvre 2007)" (416). Collaborators add to existing conversations, whereas conveners bring people together online to work on a project, unlike a webinar, which is primarily an online meeting (416). Lefebvre defines a broker as a "dynamic resource center" where people can share advice, discuss their work, and seek out new ways to reach their priority groups (Lefebvre 2013, 416). Network weavers create bridges between groups who do not usually work together, similar to the way bridging social capital facilitates the sharing of information and resources as discussed in an earlier chapter.

Typically, social media initiatives use one or two platforms that are focusing on individuals. Lefebvre emphasizes that nowadays what matters most is reaching existing groups of people in as many ways as possible, but these must be groups who are already involved in our mission (417). For example, it is much more effective to reach out to online groups who are already attempting to lose weight if your mission is to reduce obesity. These groups may not be as large as a social media site, but they will be much more likely to pay attention to what you bring to the table than people who are thinking about fantasy football or dating. Furthermore, the groups who are discussing losing weight have valuable connections to similar networks and resources.

To use social media most effectively, we must first identify how and where our priority groups are using it. According to Lefebvre (2013), besides trying to find these conversations on your own, you can use research that already exists. For both mobile and social media usage, he recommended the Pew Internet & American Life Project that conducts random surveys of Internet usage (www.pewinternet.org) and the "comScore, eMarketer, MarketingCharts, and MobileMarketer websites" (418). In terms of emergency preparedness, Lefebvre highlights the CDC's zombie apocalypse, mentioned earlier in the book, as an excellent example of the "five Es: education, engagement, entertainment, empowerment and evangelism" (429) that need to be kept in mind when using social media.

Along with the five Es, there are also five "fictions" about social media (419–424). The first is that we can reach people through a few large media sites, such as Facebook. As noted earlier, this is a fallacy because we have to first locate the sites where our priority groups are interacting, and then attune our input to their current ongoing conversations about our mutual topics. The second fiction is that we will be able to change people's behavior through social media. In fact, we need to tap into the networks of people who already want to change, such as, in our case, people with functional needs who want to be safe in their homes during emergencies. Using general media sites is not effective because, as Lefebvre emphasizes, most people who use social media do not want to interact with somebody who is trying to change their behavior. They want to talk about a specific object (e.g., cats, books, family) that they are interested in with like-minded others.

A third fiction is that people who use social media are very interactive. In fact, Lefebvre cites Forrester Research's (Bernoff 2010) definition of seven online groups along with their estimated participation. The groups and their percentages are as follows: inactives (17 percent), spectators (70 percent), joiners (59 percent), collectors (20 percent), critics (37 percent), conversationalists (33 percent), and creators (24 percent) (Lefebvre 2013, 422). This breakdown reveals that most people are not very active online. Lefebvre points out that this should help us to recognize that just having the right message will not make people stop or adopt some behavior. In fact, he states that people might need to be in one of the more active communities before they are ready to change (423).

The fourth fiction is that our priority audiences are active online. Actually, they may be, but they will be in a variety of communities and we need to find them. Lefebvre reiterates that it is our interpersonal communication within these communities that makes the most difference (423). People who are online will comment on and perhaps change what we say, which is yet another reason why one-way messaging is very unlikely to be successful. The fifth and final fiction is that we are unlikely to find our priority groups on the most popular sites (424). It is far more likely that they are searching for very specific information, as people do when they are looking for advice about a medical condition.

In 2007, Lefebvre advised that social marketers who want to engage their priority groups online should do six things: be on as many online formats as possible, encourage their audiences to prepare content, collaborate, engage social networks, and draw on social capital, and, as mentioned previously, find priority groups within larger communities and focus on the five Es noted earlier (Lefebvre 2013, 428–29). As yet, there is little research about the effectiveness of these processes, because until recently the focus has remained on standard measures of messages sent and received and the number of followers (431). These same principles should be applied to mobile technologies given their nearly universal use and similar potential for facilitating community social change (443).

Hopefully, if you have read the book thus far, you are considering beginning a social marketing program. If you are, but are hesitant about your ability to obtain the personnel and resources to do so, please read the next two sections carefully.

Start Where You Are

There is no problem at all with making small, incremental changes to your current public emergency preparedness outreach program when that is possible or as funding becomes available. Particularly in government, incremental change is the norm. It is very difficult to obtain funding for and to implement large changes, plus that would probably be disruptive for staff, clients, and citizens. Lefebvre

(2013) lists several ways that social marketing programs can be improved. First, he suggests focusing on the true priority groups, not just those we want to help. This means the groups crucial for success, which for most agencies includes the agency heads, elected officials, and other decision makers. Lefebvre states that he found this serious omission in every program he evaluated closely (306).

Second, even if you never intended to develop a brand for your program, you have one. You can discover what that brand is by listening to what people say about your organization. If you don't like what you hear, rather than developing some new message, you need to change what your clients experience so that it is representative of your mission and goals. Third, Lefebvre cautions that partnerships and collaborations are frequently ignored because, as mentioned previously, they are not priority groups (306). However, as discussed in the chapter on working with your local community, these relationships are vital to the success of social change efforts and so need to be nurtured. Fourth, goals and objectives are important, but you will be most successful if you define the behavioral changes that you want people to make very specifically (307). Finally, Lefebvre suggests taking what I have previously called a strength-based approach, and rather than focusing solely on barriers and problems, spend time envisioning aspirations and appreciating assets (2013, 307).

Small Wins

According to Weick (1984), a major impediment to solving social problems is how they are defined. When problems are defined as hugely intractable, people become overwhelmed and fearful of even attempting to begin to try and solve them. Indeed, many problems, particularly those addressed by government and nonprofits, are wickedly difficult to solve and cause what Weick called dysfunctionally high levels of arousal in us. This situation prevents us from acting at all, much less being relaxed enough to creatively search for innovative solutions.

Instead, Weick advocated the use of Alinsky's (1972) strategy of small wins. Alinsky, a community organizer, recognized that he needed to inspire hope in a demoralized neighborhood organization by having them identify and fight for an issue that was achievable. Having done this, they were enthusiastic to pursue another goal to improve their lives. Alcoholics Anonymous is another classic example of a small win. In the twelve-step program, people are not asked to pledge to never take another drink, but to stop "one day at a time." That small win is supported by other simple and accessible encouragements, such as group meetings, information, and slogans (Weick 1984, 42). Weick defines a small win as one that is "a concrete, complete, implemented outcome of moderate importance" (43). He acknowledges that in and of themselves, such wins may seem unimportant, but their cumulative effect and especially the momentum they provide for people to do more are significant.

Small wins are effective because they break apart the puzzle of a large problem in which each piece is tied to the other like a knot. The smaller piece causes less anxiety and people can see the effect of their actions quickly. Weick notes that breaking down a large problem reduces the importance of a failure ("this is no big deal"), is less demanding on people ("that's all that needs to be done"), and makes the action manageable with existing resources ("I can do at least that") (46). Even if a group does experience failure, it will not be so devastating as to prevent them from taking more action.

For our purposes, some small wins include the following:

- Obtaining a small grant or in-kind contribution to help a neighborhood association develop individual family communication plans and assistance for seniors and those with functional needs
- A partnership between a local college communication program and a county emergency management agency to develop emergency preparedness PSAs and messages for priority groups and youth
- An emergency preparedness training program on family communication for a local school sponsored by a local business
- A social marketing pilot project to train seniors and people with functional needs to take the three basic preparedness activities advocated by FEMA

Another not so small win is obtaining and using volunteers, which will be discussed next.

Volunteers in Emergency Management Agencies

Brudney and Gazley (2009) analyzed data from the 2006 National Survey of Emergency Management in County Government to attempt to determine whether county emergency management (EM) agencies that used volunteers were more prepared for emergencies. The study was implemented by the National Association of Counties and the Center for the Study of Counties at the University of Georgia's Carl Vinson Institute of Government. It appears that the survey is the only one of its kind to date and included questions about logistics, training, planning, and how prepared emergency managers felt their jurisdictions were for future emergencies (381). The researchers were able to use 407 responses from the 3,066 U.S. counties, which included representation from all Census divisions in forty-six of the fifty states (Brudney and Gazley 2009, 381).

Brudney and Gazley (2009) developed a composite scale, Overall Emergency Management Preparedness of the County, using eight variables and indices from the survey (386). These variables were: including the needs of special populations in the EM plan and arrangements to meet those needs, the comprehensiveness of the plan (e.g., an EOP, hazard mitigation plan, and/or evacuation plan), the existence of EM alerts and shelters, the most recent disaster drill, emergency agreements and funds, and access to internal and external agency and the private sector (386). The eighth variable was the extent to which each of the following groups was prepared for disasters common to each county: the majority of people, most government departments, most private businesses, most hospitals, the police, the fire department, and schools (Brudney and Gazley 2009, 386). These items were rated on a five-point scale ranging from a very great extent to no extent.

The survey included questions about volunteers in county emergency management agencies, both helping with routine duties and with nonroutine activities such as emergency planning and disaster response. This enabled Brudney and Gazley to analyze whether volunteer involvement affected county preparedness. After extensive statistical analysis controlling for the size of the county, level of risk, training, and full- or part-time status of the emergency manager, they found that counties who included volunteers in routine and nonroutine activities, such as planning and disaster response, were more prepared for emergencies than those who did not.

Furthermore, they suggested that county emergency managers consider not only involving volunteers who routinely participate in emergency planning and response, such as the American Red Cross and other Voluntary Organizations Active in Disaster (VOADs), but others as well. They identified as secondary responders those agencies that provide services to people after disasters, such as United Way agencies, churches, and mental health counselors. Tertiary responders were organizations that serve large numbers of people, such as schools and colleges, membership associations, charities, and businesses (391). Yet Brudney and Gazley made these suggestions fully aware that many emergency management agencies hesitate to involve volunteers in either routine activities or responses, because to do either is time consuming and they may question the abilities of citizen volunteers in disaster response. Other agencies may still lack the capacity to coordinate with other governments, much less to work with volunteers at any level. Nevertheless, it is clear that volunteers can be a valuable resource if managed properly. We now move to a discussion of that process.

Developing and Sustaining a Volunteer Program

The first step in beginning a volunteer program is to define what being a volunteer means in your organization. According to Macduff (1996), some of the aspects to consider are: the type of volunteerism, such as whether it is mandatory or elective, its function (e.g., administrative support), the types of tasks that will be performed (e.g., office duties, emergency assistance in the field), and the frequency of service (e.g., hours per month or as needed). Cnaan and Amrofell (1994) identified additional factors, including the demographics of the volunteers, what they donated, who benefits from the service, who manages the volunteers, and the length of their commitment. Once the volunteer description is developed the broader implementation plan can be fleshed out.

FIGURE 6.1 There is a volunteer role for everyone who wants to help others (Image courtesy of leremy/iStock)

Brudney (2010), an international expert on volunteerism, identified five challenges that must be overcome to establish and maintain successful volunteer programs. First, paid employees must accept working with volunteers. Volunteers then need to be recruited. The third challenge is insuring that the volunteers will be just as accountable to the agency as paid employees. Next, the agency needs to manage the volunteers and their activities. Finally, volunteer activities and performance must be evaluated (232). Because it is unusual for any agency to fully address these challenges, it is worthwhile to discuss each one in more detail.

Volunteers are generally considered a "free" resource that can be used on an as needed basis without the planning and oversight necessary for paid employees. Quite the opposite is true, not only because of agency liability but because of the nature of volunteerism itself. Volunteers are eager to serve but they need to know what they will be asked to do. They will want feedback and recognition for their efforts. Because they lack the incentive to stay on the job for pay, the agency needs to identify their motivation to serve and match it with opportunities that are intrinsically rewarding.

Insuring that paid employees accept volunteers is one of the primary challenges when bringing volunteers on board and one that is frequently overlooked. Brudney (2010) indicates that this problem can be overcome by having the agency's leadership and employees develop the program along with the citizens who will be volunteering. While working together, paid staff learns why volunteers are needed, what they will and will not do, and how they will be managed. This gives paid employees and future volunteers the opportunity to work out potential conflicts and begin to develop relationships with one another. Once the parties have come to consensus, formal job descriptions should be written for the volunteers to distinguish their work from that of paid staff. Doing so shows the agency's commitment to both groups and to quality service delivery.

Recruitment is the second volunteer management challenge. As Putnam (2000) and others have documented, there has been a steep decline in civic engagement in recent decades, including volunteerism. However, Brudney (2010) has identified ten ways to recruit volunteers (241–242). Because each of these methods may be useful at various points, we will briefly review them all. The first two methods are based on the fact that many people volunteer because they have been personally asked to do so. Therefore, agency staff should routinely reach out to potential volunteers and publicize volunteer opportunities. The third method is to create volunteer jobs that match what people are motivated to do. Of course, this can only be done by asking potential volunteers what their skills are and what they would like to do.

Fourth, staff should bear in mind that people frequently volunteer to build their skills or contacts in order to improve their ability to obtain paid work. Hence, positions might be created to fulfill this need. The fifth through eighth methods revolve around job redesign, including enabling people to volunteer with their family or faith community; electronic, virtual volunteering; short, time-limited, or episodic opportunities; and having a variety of times when people can serve and still be provided with equipment and reimbursement for personal expenses to do so. A ninth method is to work through umbrella volunteer organizations to obtain volunteers. In terms of emergency preparedness, Citizen Corps immediately comes to mind. Finally, Brudney suggests public recognition of volunteer service by elected officials and policymakers.

Consistent and clear oversight is necessary to insure that volunteers are accountable. Someone must be designated to monitor and coordinate volunteer activities, including performance evaluation. Although it is apparent how important this is to an agency's mission and reputation, Brudney (2010) refers to a National Association of Counties survey that revealed that less than 40 percent of counties have a volunteer coordinator (Lane and Shultz 1996, 7) but he also notes that this is more than most cities or other jurisdictions have (244). Many organizations simply add volunteer management to an employee's job description and hope for the best.

Lack of adequate supervision may increase liability for the agency and shows less regard for volunteers' needs. For this reason, a volunteer handbook listing all applicable policies is an important vehicle to maximize accountability. The guidelines in the volunteer handbook should mirror those for paid employees in applicable areas so that everyone is being held up to comparably high standards (245).

The key resource for meeting the management challenge is the volunteer job description. Brudney (2010) emphasizes that "all other aspects of volunteer management . . . emanate from the job description" (247). As noted previously, the volunteer job descriptions should be developed with broad input from senior managers, employees, and potential volunteers themselves. Employees should be asked how they might want to use volunteer help and to identify any gaps that could be filled by additional qualified personnel. The job descriptions should include the standard information; for example, the job title, description, qualifications, duties, activities, location, and benefits (249).

Another area of concern for agencies that want to implement a volunteer program will be liability. Brudney (2010) maintains that good volunteer management, as outlined thus far, can eliminate some potential problems. But he notes that while legislation such as the Volunteer Protection Act of 1997 provides some guidance, volunteers are only protected from lawsuits in civil cases, not criminal. They also must work within their area of responsibility and, of course, not commit crimes, be negligent, or endanger others (249). This is yet another reason why job descriptions are so necessary and valuable.

Evaluation is the fifth challenge that must be overcome and is the one that is most frequently overlooked. Brudney (2010) emphasizes the positive effects of volunteer evaluation while warning that neglecting it can endanger accountability. Evaluation should focus on several components of the volunteer program: the clients or beneficiaries of the volunteer program, the paid employees who work with the volunteers, and the volunteers themselves (250). Many agencies only keep records of volunteer hours and perhaps services provided. Some add a dollar value to the hours donated but almost no organizations include this contribution in their operational accounting. Nor do most agencies account for what it is costing the organization in staff time, resources, etc. to maintain the volunteer program. While it is understandable that organizations may hesitate to scrutinize and therefore possibly criticize voluntary work because it may hurt recruitment or retention, without doing so the voluntary program may stray from its alignment with the agency's mission and goals (251).

Second, in order to more fully integrate the volunteer program with normal operations, paid employees who manage or work with volunteers should have those duties included in their job descriptions. These employees should be trained to work with volunteers and have those responsibilities evaluated as well. Third, the volunteers themselves should be evaluated based on their job description. Brudney (2010) notes that these evaluations are rarely done. He goes so far as to state that not giving volunteers feedback when they have given their time and might benefit from that information "is both a breach of professional ethics and a disservice to the volunteer" (252). Furthermore, Ellis (1996) considers volunteer performance evaluation a compliment, because doing so recognizes that the person is interested in doing well and would want to improve.

Brudney suggests two final types of evaluation that are helpful in maintaining a quality volunteer program. One is to routinely determine whether the program outcomes are in line with the agency's goals and what the program itself was intended to achieve. The second type focuses on the success of the volunteer program processes themselves, such as recruitment, training, and retention, in order to identify strengths and areas for improvement (253).

Volunteers for Social Marketing for Public Emergency Preparedness

Because volunteers should be managed like paid employees, the best policy is to have someone assigned that duty. A volunteer manager would be a valuable asset to any emergency management agency to coordinate all the aspects of volunteer management; for example, Citizen Corps, MRC, and VOAD. But for our purposes, the volunteer coordinator for a social marketing program would most likely be an emergency management agency employee who was either running the program itself or working with a consultant or community volunteer who was doing so. At this point we are assuming that senior management is in full support of the social marketing program and has conveyed that support to staff.

Volunteers for a social marketing initiative could fill some of the following roles:

- Identifying the tasks that need to be done (e.g., recruitment, volunteer and citizen training, and recordkeeping) and preparing the paid agency staff for working with volunteers by polling them on their needs for the additional work created by the social marketing initiative
- Preparing job descriptions for the tasks to be completed, including all the elements described previously
- Assisting with volunteer management coordination by keeping records on volunteers, including polling them on their skills and motivation for volunteering, credentialing them, and identifying who will supervise and evaluate them
- Recruiting volunteers using the ten techniques described earlier
- Preparing a volunteer manual that parallels guidelines for paid staff
- Reporting on the social marketing initiative itself

Some tasks that are unique to the social marketing process are field research to identify priority groups, perhaps including Geographical Information System

TABLE 6.1 Sources of volunteers for a social marketing process

Citizen Corps
National Senior Corps
Youth groups (Scouts, etc.)
Social marketing professional associations and university programs
K-12 fire prevention teachers
High school service volunteers
Colleges and universities: interns, class projects, professors
Faith-based organizations
Civic organizations
Disaster survivors
Building Owners and Managers Association and similar professional associations
Business Continuity Planners
Insurance companies
Local Emergency Planning Committees
Community Development Corporations

maps, statistical analysis and longitudinal recordkeeping, developing social marketing messages and methods (the 4Ps), and community organizing to identify and increase social capital. Some sources of volunteers for these different roles are listed in Table 6.1.

Identifying and obtaining alternative sources of funding will be discussed next.

Funding

Lee and Kotler (2011) note that there are three basic types of budgeting models typically used for social marketing initiatives: affordable, competitive-parity, and objective and task (417). The affordable model is the one typically used in local government in which funding is based on what is available in the annual budget or on what has been spent on a program in the past. A competitive-parity budget is based on what another organization has spent for a comparable effort. The third budget type, objective and task, is the most commonly used. This is an incremental model that proceeds in phases. Goals and objectives are identified along with their associated tasks. The costs of the tasks are determined and the project proceeds using available funds to complete as much work as possible. This model has the advantage of allowing more time to obtain further funding from other sources through the routine budgeting process.

To reiterate the specifics of what was covered regarding the local social marketing program discussed earlier in the book, funding needs to be allocated for the 4Ps: Product, Price, Place, and Promotion. Product includes training-related materials and the training itself (i.e., in-house printing, training conducted by staff, grant writing, managing volunteers, and obtaining materials for seventy-two-hour kits via donations). Price refers to incentives or rewards (only training certificates or donated goods). The place will be where people usually meet. The only cost is mileage for trainers. Promotion will be done through existing channels. Evaluation will be done as part of training and follow-up. Additional funding will be sought through grants from the state or federal government, and from service agencies for seniors and/or people with functional needs.

Lee and Kotler (2011) identified five types of funding sources in addition to the amount budgeted for the project: government grants, foundations, media, partners, and corporations (421–427). Government funding is the most common of the five types. Regarding the second type, they noted that in 2010 there were over 88,000 active foundations in the United States that were focused on social issues (421). Kotler and Andreasen (1991) grouped these foundations into four types: family, general, corporate, and community.

The third type, advertisers and media organizations, may donate their services or allow a corporate sponsor to partner on the social marketing project and pay for marketing ads (423). In a coalition or partnership, an organization may present a social marketing message on their premises. Lee and Kotler (2011) use the example of a department of motor vehicles running a video on traffic safety in their waiting room (423). The final category is corporations. Assistance from them can take four different forms: cash grants and contributions, cause-related marketing (e.g., a percentage of sales is donated to a nonprofit), in-kind contributions, and use of their distribution channels (425–427).

Lee and Kotler (2011) suggest using the same strategies with potential funders as when working with priority groups. This means that the funders need to be segmented into those that have the best fit and potential for providing the necessary resources (427). These might include funders that you already have a relationship with, those who share your goals, or whose constituents include your priority audiences. Next, you should prepare a specific request, including the benefits to the potential funder and any concerns they might have about the project. Lee and Kotler recommend using the marketing mix four-step process that incorporates value to the funder and the benefits to your program, speaks to the funder's concerns, clearly describes how the funds will be administered, and defines how results will be measured (428). Calculating the return on the funder's investment (ROI) will be discussed next.

Return on Investment

Return on investment is calculated after the project is completed and, as Lee and Kotler (2011) explain, has three benefits. First, it provides a reason for continuing to fund the program. The ROI can also provide important information when budgeting decisions are being made, that is, programs with greater ROI may receive priority. Third, if more programs calculate ROI, comparisons can be made between them (401).

There are five steps that precede calculating the actual and final ROI (Lee and Kotler 2011, 401–402). The first step is to determine the total amount of money spent, including the value of staff time. The second step is to determine the number of behaviors that were influenced, in other words, the number of people who changed their behavior because of the intervention. The third step is to divide the amount of money identified in the first step by the number of behaviors in step two to determine the cost per behavior. The fourth step is the benefit per behavioral change. These benefits might be reduced medical costs or environmental improvements. This of course can be difficult to answer, especially because an outcome could be the result of variables besides the intervention itself. Lee and Kotler (2011) suggest at least attempting to make this calculation, being transparent about how you arrived at the figure and identifying it as a "best estimate" (402).

The actual ROI is calculated in three steps. First, multiply the number of behaviors influenced (step two) by the benefit per behavioral change (step four). This yields the gross economic benefit. The next step is to subtract the amount spent from the gross economic benefit. The remainder is the net benefit. The final rate of ROI is determined by dividing the net benefit by the total amount of money spent (step one) and multiplying that number by one hundred (Lee and Kotler 2011, 402).

Depending on the specific goals and objectives of the public emergency preparedness social marketing project in your community, some benefits might be:

- Reduced calls to 9–1–1 during emergencies
- More efficient and thorough evacuations
- More people effectively sheltered-in-place
- Reduced injuries and visits to emergency rooms
- Fewer calls to the American Red Cross "Safe and Well" program because families have communication plans, including out of state contacts
- Fewer insurance claims, because people have prepared their homes and property beforehand
- A greater percentage of the community signs up for advance alerts and warnings and knows where to obtain accurate information, reducing rumors and unsafe behaviors
- More people know the emergency procedures at their workplace, place of worship, shopping areas, etc., so they are able to take safe action and help others
- More people have an emergency plan for their commute, which helps prevent traffic jams and facilitates smoother evacuations
- More people have a seventy-two-hour kit enabling their family to be relatively comfortable and self-sustaining, allowing safety forces to focus on those most in need
- More people have alternative resources (flashlights, etc.) when utilities are out, preventing calls to utility companies and dangerous behaviors, such as the use of candles and barbeques indoors

Conclusion

Beginning and maintaining a social marketing program for public emergency preparedness requires resources that most public and private agencies have to obtain. Those resources may include additional personnel, including volunteers and people with expertise in using social media to interact with priority groups and other stakeholders. Funding may be another challenge, particularly because the use of social marketing is more common in public health and other disciplines that budget for outreach to the public. Beginning with a small pilot project and a few small wins can facilitate calculating a return on investment that can be used to justify expansion of the program based on benefits to the local community. The next chapter explores additional methods to persuade communities to become more prepared.

References

Alinsky, S.D. 1972. *Rules for radicals*. New York: Vintage.

Bernoff, J. 2010. *Introducing the new technographics*. Cambridge, MA: Forrester Research. https://www.forrester.com/Introducing+The+New+Social+Technographics/fulltext/-/E-RES56291

Brudney, J.L. 2010. Using volunteers in the workplace. In *Handbook of human resource management in government*. S.E. Condrey, ed., 231–255. San Francisco, CA: Jossey-Bass.

Brudney, J.L. and B. Gazley. 2009. Planning to be prepared: An empirical examination of the role of voluntary organizations in county government emergency planning. *Public Performance & Management Review*, 32 (3): 372–399.

Cnaan, R.A. and L. Amrofell. 1994. Mapping volunteer activity. *Nonprofit and Voluntary Sector Quarterly*, 23 (4): 335–351.

Ellis, S.J. 1996. *From the top down: The executive role in volunteer program success*. Rev. ed. Philadelphia, PA: Energize.

Kotler, P. and A. Andreasen. 1991. *Strategic marketing for nonprofit organizations*. Englewood Cliffs, NJ: Prentice Hall.

Lane, P. and C. Shultz. 1996. *Volunteerism in county government survey results*. Washington, D.C.: National Association of Counties.

Lee, N.R. and P. Kotler. 2011. *Social marketing: Influencing behaviors for good*. 4th ed. Thousand Oaks, CA: Sage.

Lefebvre, R.C. 2007. The new technology: The consumer as participant rather than target audience. *Social Marketing Quarterly*, 13:31–42.

———. 2013. *Social marketing and social change: Strategies and tools for health, well-being, and the environment*. San Francisco, CA: Jossey-Bass.

Macduff, N. 1996. *A volunteer taxonomy*. Paper presented at the annual meeting of the Association for Research on Nonprofit Organizations and Voluntary Action, New York, NY.

Putnam, R. 2000. *Bowling alone: The collapse and revival of American community*. New York: Simon & Schuster.

Weick, K. 1984. Small wins: redefining the scale of social problems. *American Psychologist*, 39 (1): 40–49.

7

EDUCATION, SOCIAL MARKETING, AND THE LAW

> In sum, the goal of both social marketing and policy advocacy is to influence behavior to deliver a positive benefit for society.
>
> *(Novelli and Workman 2011, 389)*

The foundation of this chapter is Rothschild's (1999) continuum of interventions, which spans from education through social marketing to law. Within that framework, we next explore techniques of persuasion useful to the social marketing of public emergency preparedness. The next section is a somewhat in-depth discussion of the use of behavioral economic theory, particularly by governments. By using "nudges" and default options, governments around the world are enabling their citizens to make better decisions without eliminating their freedom to maintain their less beneficial current behavior by making it illegal.

Because marketing is not something that many emergency preparedness professionals are comfortable with, the next portion of the chapter builds on the social marketing plan developed earlier in the book to craft messages for priority audiences. The chapter concludes with a discussion of the complementary nature of social marketing and policy advocacy. At that point we return to the NAS recommendations for achieving National Resilience by 2030 to see how education, social marketing, and the law might be used to expedite their adoption.

A Continuum of Interventions

Although the focus of this book has been on using social marketing to improve public emergency preparedness, it can be used in combination with other strategies.

Education through information dissemination, training, and ongoing follow-up is an important component of the process, as is, of course, social marketing. A third strategy is through a variety of coercions, some gentle and some harsh, through the use of public policy and the law.

Rothschild (1999) described this continuum of interventions, education, social marketing, and law, which can be used to manage public health and social behaviors based on the target audience's stage of change. His continuum was based on the work of three scholars (25). In *Politics and Markets*, Lindblom (1977) described three types of social control: persuasion, exchange, and authority. Next, Rothschild drew on Smith's (1996) five public health promotional techniques, "the 5Fs: facts (informational education), feelings (emotional education), facilitation (product, price and place), freebies (promotions), and force (the force of law)" (Rothschild 1999, 25). The final framework was from Hastings and Elliott's (1993) study of the use of marketing for traffic safety using the "3Es—education, environment, and enforcement."

Within the continuum, education is defined as containing messages to attempt to change behavior voluntarily, without rewards or punishment. Next, in addition to messaging, marketing employs incentives and consequences to encourage voluntary behavioral change. Finally, law attempts involuntary behavioral change through coercion and punishment (Rothschild 1999, 25). An important distinction between education and marketing is that education implies some future benefit that the recipient must seek out, whereas marketing defines the time frame for a specific benefit (26).

Due to their lack of training in marketing, many managers do not appreciate the importance of considering "the self-interest of the target, the benefits of an exchange, and the constraining nature of power and competition" (Rothschild 1999, 26). In terms of public emergency preparedness, these factors highlight why information dissemination and education have not been very successful. First, emergency preparedness professionals rarely have the time or resources to identify the self-interest of specific priority groups. Second, there is rarely an incentive or exchange. Donated preparedness items are not sufficient incentives to motivate change. Third, particularly in regions that experience few emergencies, nearly everything in daily life "competes" with making emergency preparations.

Rothschild then explains primary and selective demand, concepts used in commercial marketing. Primary demand occurs when people have already made a choice about the "product class." For example, in our case, that could be a decision to store seventy-two hours of food and water. The selective demand would then be the choice of what type of food to include. He emphasizes that whether people are making decisions about purchases or their health (or in our case, emergency preparedness), they act in their own self-interest (27). The more people perceive a benefit, the more likely they are to make a purchase or change their behavior. In addition, smaller selective demand decisions are more likely than larger primary decisions.

However, purchasing decisions differ from other behavioral decisions, because their time frame is narrow and the benefits are clear. As noted previously, when education and law are used to attempt to change public behavior, the benefits are often vague and at some unspecified time in the future. Rothschild points out that some managers who believe that they are using social marketing overlook this distinction. This is clearly shown when we are asking people to give up time and resources to prepare for an emergency that may not happen. We will discuss some ways to overcome this barrier later in the chapter.

At the opposite end of the continuum is the force of law. Although coercing people into engaging in behavior that is beneficial to them, for example, wearing seat belts, can be very effective, it is not appropriate for emergency preparedness generally. There are, however, mechanisms that can be used via public policy that automatically have people behave in a certain way, unless they take the time to 'opt out' Although not yet mandatory, some states give people the opportunity to provide an emergency contact when they renew their driver's license. These types of policies, based in behavioral economics, will be discussed later in this chapter.

Nancy R. Lee, a well-known and highly regarded social marketing expert, recently came up with a catchy way of thinking about the continuum. On the Georgetown Social Marketing listserv on August 28, 2014, she wrote that her "latest favorite way of looking at the population relative to a behavioral change was as either 'Show Me' for whom education works; 'Help Me' for whom social marketing works, needing more than words to overcome barriers and increase benefits; and [the] 'Make Me' group, being persuaded most often by laws and [the] perception of enforcement."

Rothschild cited Wiebe (1951) who said it was "difficult to sell brotherhood like soap" (Rothschild 1999, 36) because, whereas marketers can identify a specific product, social marketers sometimes cannot. Because this can also be a problem with public emergency preparedness, Table 7.1 provides some rationales for taking the three preparedness steps. I have specified safety, comfort, and peace of mind most prominently because these may be the most compelling arguments, followed by protection of property. However, these rationales have not been verified experimentally, so this is an area for further research on public emergency preparedness.

Persuasion Techniques

As has become clear thus far, rational arguments are not usually successful in getting people to adopt or reject a specific behavior. Fortunately there is a large body of research on persuasion techniques. We will review just a few concepts most pertinent to preparedness. Petrova and Cialdini (2011) recommend three strategies to overcome resistance to persuasion. These three strategies are based on a surprising finding from the research: sometimes if people are warned about something, it may actually draw them to the behavior (113). The authors provided two examples. First they noted that some people actually became more interested in viewing a program because they were cautioned about its violent content. In

TABLE 7.1 Reasons to take three Ready.gov preparedness actions

1. Be informed—for all hazards	**Reasons**
Know what to do before, during, and after an emergency	To be safe, self-sustaining, and comfortable at home versus in a shelter
Physical safety (Shelter-in-Place (SIP)/Evacuation)	Don't have to call 9-1-1 when they may not be able to respond
Family communication plan	Peace of mind knowing where loved ones are
Emergency alerts, local emergency plans for SIP/evacuation, local emergency contacts, advance alerts and warnings	To have information when rumors may abound To be able to take action based on this knowledge
Recovery safety (and mental and physical well-being)	Safety Emotional stability
2. Make a plan	**Reasons**
How to get to a safe place, contact one another, get back together, and what to do in different situations	Peace of mind knowing where loved ones are and how to get back together
Know emergency plans at work, school, church, shopping, etc.	Be knowledgeable to safely help family, friends, and others
Commuting emergency plan	Be knowledgeable to safely help family, friends, and others
3. Build a kit	**Reasons**
Three days of water, food, and supplies to stay or go	To be safe, self-sustaining, and comfortable at home if required to SIP or in a shelter if required to evacuate
Alternatives if utilities are out	Be safe and relatively comfortable Protect family, home, and property

another scenario, people who were encouraged to eat healthier foods increased their consumption of less healthy options. In terms of emergency preparedness, this implies that encouraging people to prepare may in fact have the opposite effect, leading them to not take any precautions.

Fortunately, Petrova and Cialdini's (2011) three recommendations can counter this tendency. The two examples previously mentioned describe what is termed "reactance," meaning that a person views a behavioral change as a threat to their freedom. The authors suggest that this problem can be addressed by not only advocating for the better behavior rather than denying something, but also emphasizing that people have the freedom to choose and that even a small step is valuable. They use an example of fundraising in which just adding the words, "even a penny will help" nearly doubled donations (Cialdini and Schroeder 1976). This "one small step" is reminiscent of Weick's (1984) small win research discussed earlier in the book.

A second type of resistance is skepticism. The authors propose several techniques to overcome this obstacle (114). The first is to have strong counterarguments to objections. In our case, this might be to remind people how badly they would feel if they had not taken a few minutes to develop a family communication plan only to find that they could not locate their loved ones during a crisis. Another well-researched technique is to inform people that many others are engaging in the behavior and inviting them to 'join the crowd.' Petrova and Cialdini (2011) cite an effort to encourage people to reuse their towels during a hotel stay (Goldstein, Martin, and Cialdini 2008) (114). Just letting people know that most people reused their towels increased reuse by 26 percent compared to patrons just being encouraged to help the environment by reusing their towel. A final technique is to use personal stories about undertaking the behavior. For our purposes, this may be very effective if the storyteller was a local person (or family member) who had survived an emergency because of their proactive preparedness.

The final barrier to positive change is inertia. The authors advocate two methods to move beyond this impasse (115). First, the decision-making process should be as simple as possible so that people do not freeze up when choosing between options. In addition, they should be reminded of a time when they had successfully done a similar behavior in the past. An example for emergency preparedness would be to have people recall when they had packed for a vacation or out-of-town trip.

In addition to the points made by Petrova and Cialdini earlier, Goldstein, Martin, and Cialdini's best-selling book, *Yes!: 50 Scientifically Proven Ways to Be Persuasive*, has many examples of persuasion for behavioral change gleaned from their own research and that of numerous other scientists. Several of them offer insights into how we can enhance the social marketing process for public emergency preparedness. First, similar to the towel reuse example noted previously, rather than focusing on the majority of people who do not prepare, we should be encouraging people to join the many who do.

Here is a vivid example: on December 6, 2014, the population of the United States was approximately 319 million people according to the U.S. Census Bureau (2014). We typically estimate that only one-quarter of the public has prepared for emergencies. Imagine if instead of touting that low figure, our message was "join the over 79 million Americans who know their community's emergency plan, have seventy-two hours' worth of supplies, and a family communication plan." This would be even more striking on the local level. In 2010, the county that has been the subject of this book had a population of 1,280,122 people. Our message could easily be, "join the over 320,000 people in our county who are ready for emergencies. Here's how . . ." As the authors Goldstein, Martin, and Cialdini (2008) note, this type of message is much more likely to be received favorably (13).

Another strategy is to attach a value to something we are giving away (37). For example, "the backpack and equipment that you are receiving as a newly trained CERT volunteer would have cost you seventy-five dollars. Please maintain it." The same principle could be applied to emergency preparedness trainings and other types of materials.

Yet another piece of scientifically tested advice is to ask people to take a small step before a larger one. So, for example, in our case, we might ask people to first listen to a five-minute video about emergencies in their local area and how to obtain accurate information about them before we asked them to prepare a family communication plan. According to Goldstein, Martin, and Cialdini, the ease and success of taking the small step builds "psychological momentum" to take another larger step (67).

Interestingly, simply providing a rationale using the word "because" makes people more likely to take action or, in the authors' example, to let someone move ahead of them in a line, "because I'm in a hurry" (151). In our case, the message might be, "please make a family communication plan today, because then you will always know that everyone is safe, no matter what happens."

It is unclear why preparedness messages routinely suggest having seventy-two hours' worth of supplies rather than the much easier to understand, "three days." This clearly goes against the authors' admonition to "not underestimate the power of simplicity" (161). Three days of supplies equals three gallons of water per person, three breakfasts, lunches, and dinners, plus snacks, potentially three changes of clothes, three days of pet food and supplies, etc. Urging people to have three days' versus seventy-two hours' worth of supplies is simply common sense given that people think of their routines much more frequently in daily cycles versus blocks of twenty-four hours.

In terms of messaging, the authors emphasize that rhyming messages are easier to remember, but research has shown that they are also perceived as being more truthful because they are easier to understand (165). For our purposes, it might prove very useful to create a rhyme or jingle based on "be informed, make a plan, build a kit, get involved" from Ready.gov.

Yet another powerful insight from Goldstein, Martin, and Cialdini is that people are more likely to continue a program if they feel like they have already made some progress. The authors use the example of giving people a few stamps on a card toward a car wash as they begin (171). In a controlled experiment, those people who had two of ten stamps at the outset were more likely to complete their card and get a free car wash than those who received a blank card with eight spaces at the beginning. An immediate example that comes to mind for emergency preparedness is to have people begin to build a kit by simply gathering items they already have. That way they have already completed a portion of their supply, giving them an incentive to continue to completion. Yet, there are methods that eliminate the necessity of people taking any initiative to begin a good behavior. To these compelling alternatives we now turn.

FIGURE 7.1 First aid supplies are an important component of any emergency kit (Image courtesy of Karen Sarraga/iStock)

Behavioral Economics or "the Nudge"

In another best-selling book, *Nudge: Improving Decisions about Health, Wealth and Happiness*, Thaler and Sunstein (2008) apply the principles of behavioral economics to public policy in an attempt to help people make better decisions. Thaler is a professor of behavioral science and Sunstein a professor of law. They begin the book with a simple example about improving health and nutrition of schoolchildren by paying careful attention to the order in which food choices are presented in a cafeteria. In a carefully controlled experiment, rearranging the food items resulted in increasing or decreasing consumption up to 25 percent (1). This makes the cafeteria manager, what the authors call a "choice architect," and they insist that "there is no such thing as a 'neutral' design" (3).

They define a nudge (rhymes with "judge") as "any aspect of the choice architecture that alters people's behavior in a predictable way without forbidding any options or significantly changing their economic incentives" (6). Using research from the past forty years about how people make choices, they conclude that rather than making rational choices, as economic theory suggests, people are more likely to go with the status quo. One key insight is that people are easily influenced by one another. Peer influence accounts for why teenage girls are more likely to become pregnant if their friends do, for our tendency to gain weight if our friends are obese, and for the influence of college roommates on academic performance (55).

Thaler and Sunstein used the acronym NUDGES to outline six principles of choice architecture. The first is to use iNcentives (99). In terms of our social marketing process, this means paying very careful attention to how the 4Ps, Product, Price, Place, and Promotion, are chosen. The second letter U, for understanding mapping, refers to their concept of a RECAP, which stands for Record, Evaluate, and Compare Alternative Prices (95). They suggest using a RECAP to enable people to decide between alternatives, for example, cell phone or credit card plans, by being able to review the past year of their usage and bills. In terms of emergency preparedness, a RECAP might be used to decide between levels of preparedness and include items such as the costs of living in an area prone to flooding or wildfires.

The D stands for Defaults, such as screen-saver settings that will remain fixed unless one actively changes them (87). An excellent default, not yet instituted for emergency preparedness, would be for citizens to be automatically included in local emergency alert systems via phone and or e-mail, rather than the current practice of having to inform, and then cajole people to adopt this service. See McKay (2014) for an additional four strategies. The G stands for giving feedback. The authors use the example of a laptop low battery indicator (93). A good example of feedback regarding emergency preparedness is FloodSmart.gov, the website of the National Flood Insurance Program. The site contains numerous resources, including an immediate flood-risk profile simply by entering your street address, which is followed by a list of nearby insurance agencies.

Expect error is the fifth principle of choice architecture. Although Thaler and Sunstein use the good example of the software "Forgot attachment?" to keep people from forgetting to do so, I prefer another one (92). In 2009, physician Atul Gawande wrote a groundbreaking, and lifesaving, book entitled *The Checklist Manifesto: How to Get Things Right*, in which he applied the principles of airline pilots' preflight checklists to medical procedures. He argued that even physicians forget simple things, like washing their hands before seeing patients. Gawande's principles are now used internationally, and not just for medicine. The use of checklists for emergency preparedness activities is rather common but can be easily overlooked during actual emergencies. This deficit accentuates the importance of not only preparedness but of routine drills in the home, office, in schools, and elsewhere, because as discussed earlier in the book, it is during emergencies that we can easily make errors.

The sixth and final principle of choice architecture is to structure complex choices so they are less difficult. The authors use the example of the information and prompts one receives when searching for a movie on Netflix (98). The tools on that site quickly direct the user to films they might like based on prior viewing. Emergency preparedness planning could be similarly structured. For example, as a first step, emergency managers could modify the three-day supply checklist to focus on items most needed for the emergencies common to that region and modify it by season, if applicable. Next, people could be asked to identify one meal selection for breakfast and lunch for three days, only varying the dinner item. The water supply would be consistent at three gallons per person with the addition of one or two other beverages for the duration and one snack per day. The list could specify two changes of clothes for the three days in addition to rain gear, etc. Adopting even a few of these suggestions would take much of the initial guesswork out of starting to build a kit. Keeping it simple is key.

Thaler and Sunstein (2008) offer numerous examples of how the NUDGES principles could be applied, including, for example, automatic enrollment in savings plans, decisions about credit (e.g., mortgages, credit cards, and student loans), prescription drug benefits, organ donation programs, saving energy and resources, and school choices. While they acknowledge that some people might be concerned about the methods they advocate to nudge the public toward making what they consider better choices, they offer three counterarguments (240). First, they urge people to judge their proposals on whether they actually have the potential to be beneficial. Second, because people always have an option to reject the default, the nudges have a much milder affect than, for example, banning some behavior outright. Finally, because in many cases something must be done, particularly by government, nudges are one solution to solving complex problems.

Lefebvre and Kotler (2011) note that the principles of behavioral economics are being used by governments around the world. They cited the United Kingdom's Institute of Government report, *MINDSPACE: Influencing Behaviour through Public Policy* (Dolan, Hallsworth, Halpern, King, and Vlaev 2010) on how it has effectively incorporated those principles into public policy (83). On May 25, 2014, Lefebvre wrote about a revision to that program on his blog, *On Social Marketing and Social Change*. The new model, EAST, reduced the number of components from nine (i.e., messenger, incentives, norms, defaults, salience, priming, affect, commitments, and ego) to four: easy, attractive, social, and timely. Within the EAST model (Behavioural Insights Team 2014), the easy factor refers to keeping messages simple and adopting defaults. The attractive factor suggests that presentation should draw people's attention, as should rewards and sanctions. The social component reinforces the importance of the human desire to conform, to react to the behavior of those in our networks, and that public commitments are more binding than those made individually. The last component, timely, is a reminder to contact people when they are most responsive (in our case, when an emergency appears likely), when their habits are already disturbed, and when they can most clearly see the costs and benefits of their actions. The following section describes methods of fine-tuning messages to increase their impact and effectiveness.

Crafting Messages for Priority Groups

In order to develop messages that will persuade members of the priority groups to take action, we have to know more about them than the type of group they are; for example, in our case, seniors and people with functional needs. Whether you will develop the messages in-house or have the luxury of hiring someone to do so, you will need a detailed description of that audience. Social marketers refer to this description as a creative brief. It was introduced to the discipline by Sutton, Balch, and Lefebvre (1995) when they were creating consumer-based health communication (Lefebvre 2013, 171).

The following discussion is based on Lee and Kotler's (2011, 320–321) description of the creative brief, which includes the purpose of the communication, target audience, communication objectives, positioning statement, benefits to promise, support for the promise, style and tone, and openings. When appropriate, I have noted Lefebvre's (2013) variations on these subject areas. Each section includes an application of the principle to seniors and people with functional needs. Many of the components for the creative brief were developed during the social marketing planning process in an earlier chapter.

The purpose describes what you want to accomplish with your social marketing effort. Lefebvre (2013) calls this the "social puzzle" that you want to solve (173). Referring back to step one in the plan, the purpose is increased public preparedness for all hazards by being informed, making a plan, and building a kit. This plan has two target or priority audiences (Lefebvre's term that I have used throughout the book), seniors and people with functional needs. The seniors are persons age sixty-five years of age or older who frequent senior centers, are eager to learn about emergency preparedness, and want to do so in a group setting. People with functional needs is a broad category that includes those who have diminished mobility, sight, hearing, or cognition, and those reliant on medications to accomplish the tasks of daily living. As with the seniors, the initial outreach will be to people with functional needs who are active with local service agencies and are eager to learn about emergency preparedness in a group setting.

The communication objectives are divided into three areas, what we want the participants to know, to believe, and to do. We want both groups to know what the three preparedness actions are, the benefits of preparing, and the risks of not preparing. We want them to believe that they are capable of preparing and of helping others to do so. Finally, we want them to do the three preparedness actions and maintain their information and emergency kit.

At this point, Lefebvre (2013) adds a situation analysis that "depicts the problem, its contours and possible solutions" (174). Seniors and people with functional needs in particular are keenly aware of the danger they could be in during an emergency if they were alone and unable to call for help. Preliminary research with these groups, however, revealed the resourcefulness of both groups. Seniors recalled the past when they survived other emergencies or made do with what they had on hand. People with functional needs shared how they face struggles in their day-to-day lives that force them to plan more carefully for situations that many people only face during emergencies, such as the inability to enter or exit a facility.

Lefebvre (2013) also include a section on competitors (174). By competitors they mean others who are trying to solve the problem, trying to maintain the status quo, or who are actively against a solution. In this case we are already quite familiar with groups attempting to prepare the public for emergencies. It is difficult to imagine who might not want people to be prepared or actively fight it. Our primary competitors are simply the hundreds of other pleasant activities that people would rather do than to think about emergencies, their risks, and how to prepare for them.

Returning to Lee and Kotler's (2011) framework, thus far, we have not discussed benefits to promise. Lefebvre (2013) calls this the key value (174) and notes that it should be from the participants' point of view. This component may be answered using the reasons to take preparedness actions noted earlier in this chapter. Both priority groups will be safer, more self-sustaining, and comfortable if they prepare. They will know how to contact others and their loved ones will know that they are safe. They will have accurate information that those who have not prepared may not. They will be able to help others, even if it is only by providing information. Finally, they will be more able to protect themselves, their loved ones, and property. The supports for these promises are the experiences and stories of those who have survived emergency situations.

Lee and Kotler (2011) refer to the next item in the creative brief as the style or tone, whereas Lefebvre (2013) actually calls it the brand (175). Because this is, in fact, a form of risk communication, it should not be frightening, but serious and with a focus on the simplicity and benefits of preparation. The openings are the opportunities to deliver the messages. For both groups this would be primarily at the places where they congregate but also via the media (radio, television, and the Internet) and particularly through their advocacy groups; for example, AARP and local social service agencies. A final additional component added by Lefebvre (2013) is called creative considerations. This refers to any limits on types of media, references to partners or logos that must be on materials, and, especially in our case, any cultural or accessibility criteria that should be considered (175).

Lee and Kotler (2011) caution that the choice of the messenger is extremely important, and so it should be given careful consideration. They state that there are six possibilities, a "sole sponsor, partners, spokespersons, endorsements, midstream audiences, [and a] mascot" (326). Each has advantages but also may have challenges because of the potential to alienate some people or become attached to a group or person that subsequently incurs some bad publicity. Midstream audiences are those closely involved with the priority group. For example, AARP is a midstream audience for seniors. Of course, more than one option is also a possibility, such as having a partner, a spokesperson, and a mascot.

In addition to the persuasion advice given previously in this chapter, including focusing on positive behavior and 'joining the crowd,' keeping the message simple and the action doable, and making messages easy to understand and remember (possibly with a rhyme), Lee and Kotler (2011) offer other recommendations. Those most pertinent to emergency preparedness include emphasizing the benefits of the action, possibly using a question, and the use of personal stories and examples (330–338). Any messages should be pretested on potential audiences to obtain their input on whether the messages will achieve their desired objectives (343–344). With a firmer foundation on developing a successful social marketing campaign, we now move into the realm of politics and law with a discussion of policy advocacy.

Social Marketing and Policy Advocacy

In their explanation of the founding and operation of the nonprofit AARP (formerly known as the American Association of Retired Persons), Novelli and Workman (2011) say that social marketing is not "broad enough" because "it doesn't fully incorporate policy advocacy, for example, and only recently began to incorporate legal sanctions" (377). AARP advocates for positive social change by influencing members to make changes that will improve their lives as they age and by advocating for policy changes that will improve the environment for making behavioral changes so that healthy behavior is what is expected rather than something special (Novelli and Workman 2011, 377).

The authors note the inclusion of three social marketing objectives within the U.S. Department of Health and Human Services' Healthy People 2020 initiative as evidence of the wide acceptance of social marketing as a large-scale behavioral change process. Those objectives are listed in Table 7. 2. The objectives fall under the category of health communication and health information technology (HC/HIT). Healthy People 2020 is a ten-year program for improving the health of the American public by reducing chronic preventable conditions, including heart disease, diabetes, and cancer.

At AARP they have used a combination of approaches via their three organizations: The parent organization, AARP, is the nonprofit advocacy organization that has over forty million members, aged fifty and older (380). The AARP Foundation is a charitable organization providing services such as free tax preparation

TABLE 7.2 Healthy People 2020—Three objectives using social marketing

HC/HIT-13.1 (Developmental) Increase the proportion of state health departments that report using social marketing in health promotion and disease prevention programs
HC/HIT-13.2 (Developmental) Increase the proportion of schools of public health and accredited master of public health (MPH) programs that offer one or more courses in social marketing
HC/HIT-13.3 (Developmental) Increase the proportion of schools of public health and accredited MPH programs that offer workforce development activities in social marketing for public health practitioners

Source: U.S. Department of Health and Human Services, Office of Disease Prevention and Health Promotion, Healthy People 2020.

and legal services. The third organization is AARP Services, Inc., which provides products and services to AARP members that may not otherwise be available. Novelli and Workman note that these organizations provide their "triple bottom line" of "social impact, member value and revenues to reinvest in the enterprise" (380). They have found that social marketing is most effective when it is used in combination with other approaches and that was echoed by Andreasen (2002), who described social marketing as complementary versus competitive.

To conclude this chapter, we will briefly return to the National Academies' (2012) six recommendations, discussed at the beginning of the book, to see where we might use education, social marketing, or the law, alone or in combinations, to move forward more effectively toward the 2030 resilience goal. Education is a component in item one of the six recommendations, which advocates the adoption of National Resilience as a "guiding principle" in federal programs at all levels (209). Recommendation two has three components. The first assigns DHS the task of developing a structural resilience framework of codes, standards, and guidelines. This seems to primarily be an educational effort. Social marketing, in particular the use of defaults, and laws might be used to build resilience in the other two components. One calls for coordination between public and private community stakeholders to reduce risks via building codes and standards and insurance based on local hazards (212). The other component has an educational mandate, to raise awareness of local hazards, but also to use risk-based insurance fees and reward compliance. Defaults and local legislation could be used to enhance the implementation of both.

Recommendation three proposed that the National Science and Technology Council in the White House develop a national disaster database of losses and their economic impact in order to build models of risk and social vulnerability (214). This also appears to be an educational activity, but it might be possible to have some of this information be "defaulted" to the national database via, for example, tax forms and insurance claims. In the fourth recommendation, DHS and other stakeholders were charged with developing a baseline National Resilience Scorecard to include recovery capability, social capital, structural resilience, business continuity, and protection of people with special needs. This massive informational effort seems to be educational but is also the type of "market" research that could be used for social marketing type default options and legislation.

DHS and the Department of Health and Human Services are charged with working together to create community resilience coalitions (216). This effort has the potential to strongly support social marketing efforts for community emergency preparedness, which by their nature are local. The final recommendation encourages all federal agencies to work together to promote National Resilience in all of their work (217). This is neither an educational nor legal process, so it seems to fall more into the community building and collaboration characteristic of the social marketing processes.

Using Rothschild's framework, social marketing should be more extensively used as a bridge between education (with little or no behavioral change) and law (forcing people to act, perhaps against their will) because it is based on an exchange in which people see the benefit of adopting new behaviors. The resilience guidelines go beyond asking citizens to "be informed, make a plan, and build a kit" to tide them over through the first three days after a disaster. They require much broader changes, such as where people choose to live and how they evaluate and respond to risks. Government has many roles; but as the NAS emphasized, much of the action leading to resilience must be taken in local communities. It is at that level that social marketing can help people improve their lives through better decision making.

Conclusion

This chapter began with an exploration of Rothschild's continuum of interventions for changing behavior, and then applied those methods to public emergency preparedness. That groundwork was followed by a brief explanation of the principles of behavioral economics and their use by governments to "nudge" the public to make better choices. Next, we used some pertinent persuasion techniques to craft messages for the two priority groups discussed throughout the book, seniors and people with functional needs. The chapter closed with some suggestions for how education, social marketing, and the law might be used to expedite the achievement of the National Academy of Sciences' recommendations for National Resilience, where the book began.

References

Andreasen, A.R. 2002. Marketing social marketing in the social change marketplace. *Journal of Public Policy & Marketing*, 21 (1): 3–13.

Behavioural Insights Team. 2014. *EAST: Behavioural insight framework*. London: Behavioural Insights Ltd.

Cialdini, R.B. and D. Schroeder. 1976. Increasing compliance by legitimizing paltry contributions: When even a penny helps. *Journal of Personality and Social Psychology*, 34: 599–604.

Dolan, P., M. Hallsworth, D. Halpern, D. King, and I. Vlaev. 2010. *MINDSPACE: Improving behaviour through public policy*. London: Institute of Government. http://www.instituteforgovernment.org.uk/our-work/better-policy-making/mindspace-behavioural-economics.

Gawande, A. 2009. *The checklist manifesto: How to get things right*. New York: Picador.

Goldstein, N.J., S.J. Martin, and R.B. Cialdini. 2008. *Yes!: 50 scientifically proven ways to be persuasive*. New York: Free Press.

Hastings, G.B. and B. Elliott, 1993. Social marketing practice in traffic safety. In *Marketing of Traffic Safety*, Chapter 3, 35–53. Paris, France: Organisation for Economic Co-Operation and Development.

Lee, N.R. and P. Kotler. 2011 *Social marketing: Influencing behaviors for good*. 4th ed. Thousand Oaks, CA: Sage.

Lefebvre, R.C. 2013. *Social marketing and social change: Strategies and tools for health, well-being, and the environment*. San Francisco, CA: Jossey-Bass.

Lefebvre, R.C., and P. Kotler. 2011. Design thinking, demarketing and behavioral economics: Fostering interdisciplinary growth in social marketing. In *The SAGE Handbook of Social Marketing*. G. Hastings, K. Angus, and C. Bryant, eds., 80–94. Thousand Oaks, CA: Sage.

Lindblom, C.E. 1977. *Politics and markets*. New York: Basic Books.

McKay, J. 2014. 5 strategies for getting the public to sign up for emergency alerts. *Emergency Management*, 3 October.

National Academy of Sciences. 2012. *Disaster Resilience: A National Imperative*. Washington, DC: National Academies.

Novelli, W.D. and B. Workman. 2011. Social marketing and advocacy. In *The SAGE Handbook of Social Marketing*. G. Hastings, K. Angus, and C. Bryant, eds., 377–390. Thousand Oaks, CA: Sage.

Petrova, P.K. and R.B. Cialdini. 2011. New approaches toward resistance to persuasion. In *The SAGE Handbook of Social Marketing*. G. Hastings, K. Angus, and C. Bryant, eds., 107–122. Thousand Oaks, CA: Sage.

Rothschild, M.L. 1999. Carrots, sticks, and promises: A conceptual framework for the management of public health and social issue behaviors. *Journal of Marketing*, 63: 24–37.

Smith, W. 1996. *Marketing and public health applying tested techniques to promote public health activities*. Paper presented to Promoting Public Health in an Era of Change, Agency for Policy and Research, User Liaison Program, Towson, MD.

Sutton, S.M., G.I. Balch, and Lefebvre, R.C. 1995. Strategic questions for consumer-based health communications. *Public Health Reports*, 110: 725–733.

Thaler, R.H. and C.R. Sunstein. 2008. *Nudge: Improving decisions about health, wealth and happiness*. New York: Penguin.

U.S. Census Bureau. 2014. "U.S. and world population clock." http://www.census.gov/popclock/.

U.S. Department of Health and Human Services, Office of Disease Prevention and Health Promotion. Healthy People 2020. Accessed December 15, 2014. https://www.healthypeople.gov/2020/topics-objectives/topic/health-communication-and-health-information-technology/objectives.

Weick, K. 1984. Small wins: Redefining the scale of social problems. *American Psychologist*, 39(1): 40–49.

Wiebe, G.D. 1951. Merchandising commodities and citizenship on television. *Public Opinion Quarterly*, 15 (Winter): 679–691.

CONCLUSION

> While citizens expect the federal government to come to their rescue when disaster strikes, it is the capacity of the local community that determines how well a community survives and how quickly life is restored to normal.
>
> *(Roberts, Ward, and Wamsley 2012, 243)*

Social marketing is a new way of thinking about preparing the public for emergencies. I have attempted to provide enough breadth and depth about the process to give the reader some confidence that it could actually be effective, even with limited funding. We have covered quite a bit of ground since the start of the book. Here are some highlights from the journey.

Chapter Highlights

This book began with a focus on the National Academy of Sciences' recommendations for achieving National Resilience by 2013. Those lofty goals were followed by a brief review of the literature on our state of public preparedness. Next, individual behavioral change theories were explained to distinguish social marketing from the methods currently being used by most emergency managers to attempt to encourage the public to prepare.

That background was followed by a brief history of social marketing and an in-depth discussion of its process, including a few applications to public emergency preparedness. The next chapter was quite different than any other in the book. We explored the opinions of seventeen emergency preparedness professionals about why people don't prepare and what could be done about that. They identified

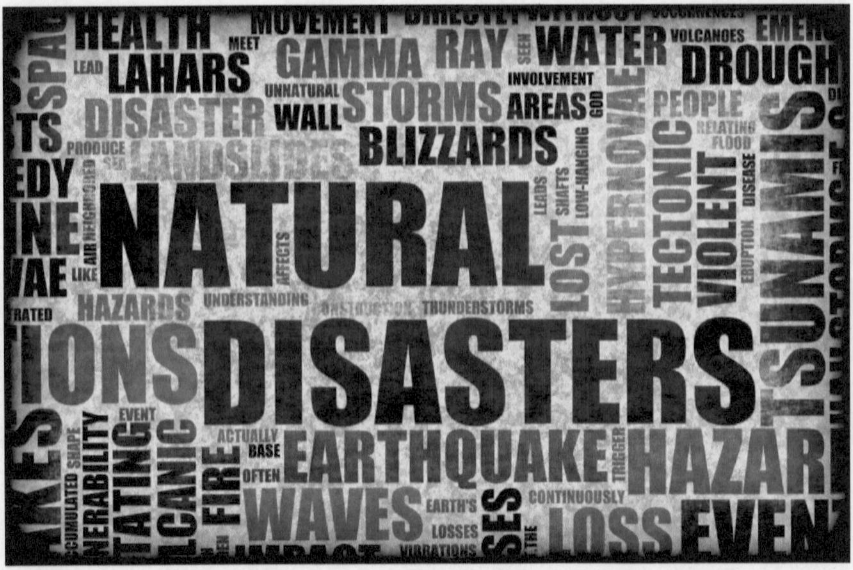

FIGURE 8.1 National Resilience requires preparation for all types of disasters (Image courtesy of Kheng ho Toh/Hemera/Thinkstock)

groups of people who were at highest risk, most ready for training, and most critical to the success of a training program. Their input revealed that, indeed, children were at greater risk, but also that training *all* children should be a priority, a given.

Seniors and people with functional needs were two other groups that were frequently mentioned. Subsequent to the interviews with the seventeen emergency managers, three focus groups were held, two with seniors and one with people with functional needs. The dialogue in those groups was formative research that identified a wide range of preparedness within both groups, but also a common desire to learn about and prepare for emergencies and to do so with their friends.

The following chapter was also unlike the rest of the book. It was based on Lee and Kotler's (2011) social marketing planning worksheets that I was able to use with Nancy Lee's kind support and the publisher's permission. With their worksheets as a framework, I fleshed out a social marketing plan for the two priority groups in my local community. The essential goal was to have a greater number of people complete the three basic preparedness actions—be informed, make a plan, and build a kit—but also to maintain them over time.

Although the nuts and bolts of a social marketing plan are critical to achieving the mission, it is working directly with people in your local community that is most important. For that reason, I included a short theoretical discussion of social capital and networks, in particular because the latter is so often confused with social media. The importance of social networks for disaster recovery is an emerging area of research, so I included a detailed discussion of some findings. There is also a brief introduction to using a strength-based approach to working with people, rather than focusing on failures and problems. That theme continues through the final chapters.

Many government agencies face a continuing lack of resources, even for routine operations. This is particularly true when the agency's mission is seen as lacking urgency, a situation that some emergency management agencies confront, particularly in regions that do not experience large-scale public emergencies on a routine basis. A chapter on alternatives to traditional resources addresses this problem, because proposing the adoption of a social marketing initiative without a plan for staffing and funding it would most likely be dismissed.

The chapter on alternatives to traditional resources was my attempt to fill in some gaps by offering guidelines on the effective use of social media (i.e., Internet communities), creating and sustaining a volunteer program, obtaining funding, and calculating a return on investment. Although relatively few government agencies maintain a volunteer program, there is a detailed discussion of how to develop one, particularly in emergency management agencies, and specifically for carrying out a social marketing initiative, such as the one described in the book.

The final chapter explains a continuum of interventions to encourage people to make better decisions that range from basic education through social marketing to the rule of law. That was followed by an overview of persuasion techniques as applied to emergency preparedness. The relatively new field of behavioral economics was introduced next via the concept of default decisions. These policies facilitate better decisions (depending on your point of view), by making the healthier or more financially sound option the status quo, forcing people to take the time to opt out vs opt in. The chapter also offered some guidance on social marketing and policy advocacy. This final chapter closed by returning to the National Academy of Sciences' recommendations for achieving National Resilience by 2030 and how social marketing could be used to accomplish those goals.

Limitations

The book has several limitations, which the discriminating reader, and especially practicing emergency managers, will no doubt identify. The primary and most serious limitation is that there is no empirical evidence that social marketing for public emergency preparedness 'works.' Based on the use of social marketing in other disciplines, I have speculated that it will work, particularly when a priority group, such as seniors or people with functional needs, is eager to learn. Second, the book only provides a framework for managers to complete based on the needs of their community. If social marketing is adopted by some emergency management agencies, I foresee that it will be in areas more at risk for emergencies and disasters than my Midwestern county, the setting for the case study discussed throughout the book.

The third limitation is the research itself. Not only is what I have prepared not generalizable, but it is also not complete. I would have greatly preferred to undertake full research studies but felt it was most important to present the social marketing process to the emergency management community as soon as possible, given the increase in natural and anthropogenic disasters.

A Final Reflection

In the course of writing this book, the world witnessed the devastation and panic caused by an epidemic of the Ebola virus. Certainly, in the United States, it was a wake-up call among public health and emergency management professionals about our state of preparedness. As I was writing, I continued to reflect on the difference a prepared citizenry could make when the next crisis occurs. I hope that this book is a step in that direction. Obviously, I am convinced that working closely with the people in our communities to help them prepare for emergencies is more effective than the traditional emergency management functions of logistical planning, reporting, and information distribution. Combining these important functions can make our communities safer and more resilient.

References

Lee, N.R. and P. Kotler. 2011. *Social marketing: Influencing behaviors for good*. 4th ed Thousand Oaks, CA: Sage.

Roberts, P., R. Ward, and G. Wamsley. 2012. From a painful past to an uncertain future. In *Emergency management: The American experience 1900–2010*. 2d ed. C.B. Rubin, ed., 237–245. Boca Raton, FL: CRC Press.

ABOUT THE AUTHOR

Nancy Meyer-Emerick, PhD, is an Associate Professor of Public Administration at the Levin College of Urban Affairs at Cleveland State University. Prior to joining the faculty there in 1999, she worked in local government in Florida for many years, primarily in public works and environmental health. In 2007, she completed a Masters in Positive Organizational Development and Change at Case Western Reserve University. As a 2009 Fulbright scholar at the University of Prešov in the Slovak Republic, she taught Organizational Behavior and Crisis Management. She is a Red Cross Disaster Action Team volunteer and a member of her local CERT.

Dr. Meyer-Emerick's research interests include organizational behavior, emergency preparedness, and environmental health. Her first book, *The Violence Against Women Act of 1994: An Analysis of Intent and Perception*, was published in 2001. Other work has been published in *Administration & Society*, the *International Journal of Organization Theory and Behavior, Nonprofit Management and Leadership, Public Administration and Development, Administrative Theory & Praxis*, the *Journal of the American Water Works Association*, the *Journal of Management Education*, and the *Journal of Homeland Security and Emergency Management.*

INDEX